ISSN 0097-0905

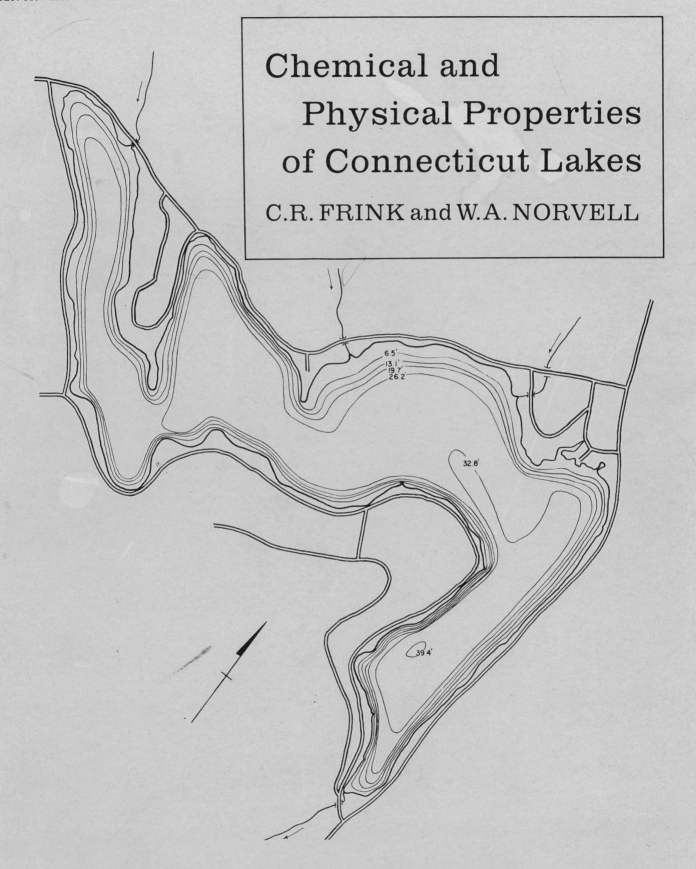

Chemical and Physical Properties of Connecticut Lakes

C.R. FRINK and W.A. NORVELL

THE CONNECTICUT AGRICULTURAL EXPERIMENT STATION

New Haven   Bulletin 817   April 1984

Pope Pius XII Library
Saint Joseph College
West Hartford, CT 06117

TABLE OF CONTENTS

Introduction	1
Methods	1
Results and Discussion	3
Major Ions	3
Trophic Status	3
Chlorophyll-a and Nutrients	5
Transparency and Chlorophyll-a	6
Changes in Lakes During a Third of a Century	7
Land Use and Lake Phosphorus	8
Summary	9
Acknowledgments	9
References	10

Appendix I. Map with names and locations of lakes.

Appendix II. Selected physical and chemical properties of 70 lakes.

Appendix III. Detailed chemical analyses of 70 lakes, 1973-1980.

Appendix IV. Bathymetric data and descriptive narrative of each lake.

Chemical and Physical Properties of Connecticut Lakes

C.R. FRINK and W.A. NORVELL

We have studied many of Connecticut's lakes and ponds during the last decade, probing their sediments, sampling their weeds and algae, and analyzing their waters for plant nutrients and other chemicals. The chemistry and fertility of 23 Connecticut lakes were reported earlier by Norvell and Frink (1975), who found that many of our lakes were undergoing accelerated eutrophication.

Simply stated, eutrophication is an increase in available plant nutrients accompanied by an increase in growth of aquatic plants. Eutrophic lakes are usually rich in plant nutrients, laden with algae and other aquatic plants, deficient in oxygen near the bottom, opaque, and frequently shallow. At the other extreme are oligotrophic lakes, which are usually poor in plant nutrients, poor in aquatic plants, well aerated at most depths, clear, and frequently deep.

In 1979-1980, the chemistry and trophic states of 47 additional lakes were determined in cooperation with the Water Compliance Unit of the Connecticut Department of Environmental Protection as required by Section 314 of the Clean Water Act. This report summarizes the analyses of all 70 lakes and interprets the results in terms of nutrient status and consequent growth of weeds and algae.

Earlier, Norvell, Frink and Hill (1979) related land use in the watersheds of 33 lakes to the concentration of phosphorus in their waters. The utility of that predictive model can now be tested on the entire set of 70 lakes. For 35 lakes, comparable chemical analyses are available for 1937-1939, making it possible to examine changes that may have accompanied changes in the environment.

METHODS

Timing and techniques of water sampling depend on the changes occurring in lakes during the year. As the ice melts in the spring, the temperature of the surface water increases slightly and its density reaches a maximum at about 4° C (39° F). This slightly denser water sinks, causing the lake to mix, and its physical and chemical properties become reasonably uniform from top to bottom. At this time, known as spring overturn, representative samples of the water column were collected from each of the 47 lakes with a weighted plastic tube that sampled the entire depth of water. These depth-integrated samples were collected at two sites in the deepest basins of each lake.

As the season progresses, the surface waters become warmer and less dense, while the cooler waters remain near the bottom. By midsummer, lakes of sufficient depth develop two pronounced layers of water separated by a rather sharp change in temperature (and density) known as the thermocline. The upper layer is known as the epilimnion and contains most of the algae and other aquatic life. The lower layer, known as the hypolimnion, is often depleted of oxygen but enriched in nutrients released from the bottom sediments (Frink, 1967, 1969).

During July and August, depth-integrated samples of the epilimnion were collected at two or more sites in the deepest basins of each of the 47 lakes. Dissolved oxygen content and temperature of the entire water column were measured at these sites with a probe lowered into the water. Samples of water below the epilimnion were collected with a Kemmerer sampler, which samples about 18 inches of the water column at any desired depth. A sample from the thermocline and one or more samples from the hypolimnion were collected in deep lakes. The transparency of the lakes during the summer was measured by lowering a standard white Secchi disk into the water and noting the depth where it disappeared from view. The approximate extent and composition of major weed beds were estimated and sketched on bathymetric maps.

Chemical analyses were performed as described by Norvell and Frink (1975) with minor modifications speci-

Table 1. Total P, total N, chlorophyll-a, and transparency for 70 lakes, ranked by total P averaged for spring and summer. Field estimates of weed problems are given for 47 lakes included in the 1979 and 1980 surveys.

Lake	Spring Total P	Spring Total N	Summer Total P	Summer Total N	Chl-a	Secchi Depth	Aquatic Weeds Area[1]/Density[2]
	ppb					m	
Bashan	8	205	5	193	2	5.5	small/moderate
Mashapaug	9	193	6	325	1	8.2	small/sparse
Highland	11	280	5	238	2	6.0	intermediate/moderate
Billings	8	184	8	320	3	4.5	large/dense
West Hill	9	262	7	245	2	7.0	
Beach	12	244	5	200	1	7.2	small/sparse
Alexander	10	308	8	405	1	7.2	
Uncas	10	145	10	254	2	5.4	small/moderate
Bigelow	9	155	11	281	9	2.5	intermediate/moderate
Mt. Tom	10	456	11	375	5	4.5	intermediate/moderate
Rogers	10	340	13	410	4	4.0	intermediate/moderate
Long	11	298	13	640	3	4.8	
Crystal	12	373	12	363	8	4.0	small/sparse
West Side	14	660	11	530	3	4.0	large/moderate
Waumgumbaug	16	424	10	416	2	6.1	small/sparse
Wyassup	16	250	10	482	3	4.3	large/moderate
Norwich	15	217	12	295	3	3.0	small/moderate
Gardner	14	472	13	595	7	4.1	
Black (Woodstock)	12	239	15	500	11	3.0	intermediate/moderate
Dodge	15	779	12	400	2	4.0	small/sparse
Powers	18	190	10	280	3	3.2[3]	small/dense
Burr	11	251	17	498	8	2.7	small/moderate
Columbia	18	259	10	263	3	5.0	small/sparse[4]
Candlewood	13	358	15	436	5	5.2	
Shenipsit	18	434	11	495	6	4.0	
Pataganset	15	388	15	680	14	2.8	
Quaddick	18	246	12	461	4	2.3	large/dense
Quassapaug	16	426	14	460	3	6.8	
Cedar (Chester)	14	413	17	285	9	4.0	intermediate/moderate
Pachaug	15	676	16	420	2	3.5	intermediate/moderate
Squantz	16	361	16	280	6	3.4	intermediate/dense
Glasgo	13	357	19	396	9	2.0	
Lower Bolton	19	358	15	869	13	2.3	small/sparse[4]
Cream Hill	16	400	18	530	4	5.5	intermediate/moderate
Little	13	644	23	365	4	3.0[3]	large/dense
East Twin	20	425	16	510	2	5.3	
Gorton	17	786	19	475	6	2.7[3]	large/dense
Terramuggus	22	453	14	590	2	6.0	
Hayward	23	370	15	370	8	3.3	
Amos	23	515	16	339	4	3.7	large/dense
Quonnipaug	27	251	12	255	5	4.4	intermediate/dense
Hitchcock	17	580	22	538	12	2.0	extensive/moderate
Black (Meriden)	25	500	17	550	10	2.5	extensive/dense
Tyler	23	630	19	569	7	3.8	large/moderate
Pocotopaug	17	510	25	420	7	3.6	
Taunton	23	370	22	690	6	3.3	
Mudge	27	553	19	483	4	4.1	
Waramaug	27	473	24	635	11	2.7	
Middle Bolton	28	470	24	757	12	2.5	small/sparse[4]
Moodus	33	345	22	519	9	2.0	large/dense
Wononscopomuc	44	692	14	510	2	7.8	
Kenosia	38	770	20	508	17	1.8	large/dense
Ball	37	888	23	544	3	2.5	intermediate/moderate
Beseck	26	550	34	647	18	2.8	
Mamanasco	27	598	34	575	11	1.8	large/dense
Bantam	26	493	35	893	31	1.8	
Winnemaug	32	1003	32	1020	24	1.3	intermediate/dense
Long Meadow	38	586	29	611	9	1.3	large/dense
Batterson Park	40	508	31	933	35	1.5	intermediate/dense
Roseland	33	953	38	935	20	2.8	
Wononpakook	43	576	28	670	15	2.1	intermediate/dense
Linsley	51	1325	28	520	5	3.5	
1860 Reservoir	59	1158	32	1260	7	1.0[3]	extensive/dense
Housatonic	41	747	51	700	22	2.0	intermediate/dense
Eagleville	34	533	60	729	14	1.5	large/dense
Cedar (No. Branford)	47	1543	71	1830	64	0.9	
Lillinonah	61	764	56	933	38	1.9	
Silver	21	400	107	1100	23	1.8	extensive/dense
Zoar	68	732	66	906	54	1.9	
North Farms	54	610	157	1375	57	1.0	extensive/dense

[1] Areal extent categories and approximate percentage of lake area: small (0-5%), intermediate (5-15%), large (15-30%), extensive (30-100%).

[2] Density of weed growth categories: sparse, moderate, dense.

[3] Secchi depth limited by depth of water in these shallow lakes.

[4] Weed growth altered by winter drawdown.

fied by the U.S. Environmental Protection Agency. Detailed information on the chemical methods is available on request.

Land use in the watersheds of all the lakes was mapped by the Connecticut Regional Planning Agencies in 1977 as part of the Connecticut Areawide Waste Treatment Management Planning Program established under Section 208 of the Clean Water Act. Fifteen categories of land use were identified (Connecticut Department of Environmental Protection, 1980). We combined these as nearly as possible into the three categories of urban, agricultural and wooded land used by Norvell, Frink and Hill (1979). Bathymetric data were obtained from the Fishery Survey of the Lakes and Ponds of Connecticut (Connecticut Board of Fisheries and Game, 1959). The area of the watershed draining into each lake, which includes the lake itself, was taken from a compilation by Thomas (1972). Other general information such as topography of the watershed and physical access to the lake was obtained from data on file at the Connecticut Department of Environmental Protection.

RESULTS AND DISCUSSION

Interpretation of the results for the 70 lakes generally follows that presented earlier for 23 of the lakes (Norvell and Frink, 1975). Table 1 contains results for total phosphorus (P) and total nitrogen (N) at spring overturn, and total P, total N, chlorophyll-a and transparency during July and August. All samples were analyzed for soluble and total nutrients, but our experience shows that the total concentration of nutrients is the more useful indicator of trophic status. Table 1 also contains estimates in the field of the extent and density of aquatic weeds in lakes sampled in 1979 and 1980. The names and locations of all the lakes are given in Appendix I. Appendix II summarizes morphometric characteristics including surface area of the lake, the ratio of watershed to lake area, mean depth, and maximum depth, as well as summarizing major ion chemistry. Complete analyses for nutrients, transparency and alkalinity are contained in Appendix III. A narrative about each lake, including available bathymetric data, is contained in Appendix IV.

Major Ions

Most of the lakes are fairly dilute bicarbonate waters. Using the categories of Brooks and Deevey (1963), 24 of the lakes are classified as extremely soft waters with bicarbonate < 0.17 meq/l; 15 are medium-hard with bicarbonate between 0.46 and 1.38 meq/l; and 9 are hard waters with bicarbonate > 1.38 meq/l. The extremely soft and soft water lakes are generally located in the eastern or western crystalline highlands, while the others are located either in the central or coastal lowlands or in watersheds with till or bedrock bearing limestone.

Norvell and Frink (1975) had observed anomalously high concentrations of Na in Linsley and Cedar Pond in North Branford. The source was apparently not domestic uses of salt or marine spray because Cl concentrations were not high. The chemical, mineralogical and chronological evidence suggest that most of the Na derives from the extension of trap rock quarries into the watersheds of these lakes (Norvell, 1977).

Trophic Status

The trophic data in Table 1 are ranked by the average total phosphorus for spring and summer. This provides a convenient ranking of the lakes since phosphorus appears most likely to be limiting the growth of algae in lakes in New England (Norvell and Frink, 1975).

Because of the large number of lakes and the wide diversity among mesotrophic lakes, the usual three category system (oligotrophic, mesotrophic, and eutrophic) was subdivided into six categories (oligotrophic, oligo-mesotrophic, mesotrophic, meso-eutrophic, eutrophic, and highly eutrophic) to allow better discrimination among lakes of differing fertility and productivity. Table 2 presents the criteria used to estimate the trophic status of each lake based on data summarized in Table 1 and detailed data in the appendix of this report.

Figure 1 shows a grouping of the lakes similar in nutrients, chlorophyll-a, and transparency obtained by cluster analysis (Dixon et al., 1981, Program BMDP2M). The 70 lakes were clustered according to six characteristics: spring P, spring N, summer P, summer N, summer chlorophyll-a, and summer Secchi depth. At the level of similarity selected for best discrimination, six groups formed that correspond to oligotrophic, oligo-mesotrophic, mesotrophic, meso-eutrophic, eutrophic, and highly eutrophic lakes. A few lakes did not cluster at the level of similarity chosen, and their assignment to the appropriate group is shown by a dotted line. Although cluster analysis is thought to provide an objective grouping of similar cases, our experience shows that assignment to groups and the order within groups depends on the variables chosen for clustering. For example, a logarithmic transformation of the data, which emphasizes relative rather than absolute differences, produced a cluster with many lakes assigned to groups different from those in Figure 1. Although others might choose to classify a particular lake differently, we feel our classifications are reasonable.

The calculation of an index of trophic state has been suggested as an alternative to dividing the trophic continuum into various classes (Carlson, 1977; Baker et al., 1981). The usual indexes are based on transparency, chlorophyll, or phosphorus. Calculations of the index as described by Carlson (1977) did not materially change the

Table 2. Total P, total N, chlorophyll-a, and Secchi depth criteria used to assign lakes to six trophic categories.

Category	Total P	Total N	Summer Chlorophyll-a	Summer Secchi Depth[1]
	ppb	ppb	ppb	m
Oligotrophic	0-10	0-200	0-2	6+
Oligo-mesotrophic	10-15	200-300	2-5	4-6
Mesotrophic	15-25	300-500	5-10	3-4
Meso-eutrophic	25-30	500-600	10-15	2-3
Eutrophic	30-50	600-1000	15-30	1-2
Highly Eutrophic	50+	1000+	30+	0-1

[1]For lightly colored lakes with 5 to 15 ppm Pt units of color.

Figure 1. Aggregation of lakes into trophic groups by cluster analysis based on spring and summer nutrient concentrations, summer chlorophyll-a concentrations and transparency.

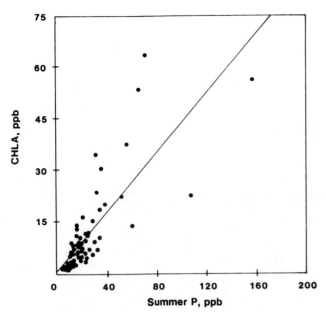

Figure 2A. Relation between summer chlorophyll-a concentrations and total phosphorus during the summer in 70 lakes on a linear scale.

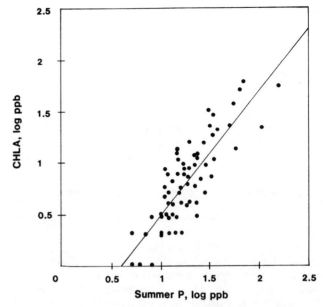

Figure 2B. Relation between summer chlorophyll-a concentrations and total phosphorus during the summer in 70 lakes on a logarithmic scale.

ranking based on the average phosphorus content shown in Table 1.

The requirements of this study included the identification of lakes where efforts to maintain or improve them might be most effective. Thus, the cluster analysis identified eutrophic and highly eutrophic lakes where elimination of sewage discharge or in-lake restoration techniques might produce the most evident results. Similarly, the analysis reveals oligo-mesotrophic and oligotrophic lakes where watershed management practices might be most effective in preventing their deterioration. However, efforts to categorize the trophic status of lakes should not conceal their existence as a continuum in the real world.

Chlorophyll-a and Nutrients

Chlorophyll-a in the surface water generally indicates the biomass of algae. In most lakes in the Northeast, phosphorus tends to limit the growth of algae unless the lake receives much effluent from sewage plants or other direct discharges of phosphorus. In such lakes with abundant phosphorus, nitrogen may become limiting, although many blue-green algae can utilize atmospheric nitrogen in a manner similar to leguminous plants.

The relation between chlorophyll-a (CHLA) and total summer phosphorus (TP) in our 70 lakes is shown in Figure 2. Figure 2A shows the relation on a linear scale according to:

$$CHLA = 0.374 + 0.431 (TP)$$

with coefficient of determination $r^2 = 0.63$. Thus, this relation accounts for 63% of the variability in chlorophyll and phosphorus.

Many investigators have chosen to transform the data to a logarithmic scale as shown in Figure 2B. In this case, the relation is:

$$Log\ CHLA = 0.711 + 1.21\ Log\ (TP)$$

with a slight improvement in the coefficient of determination to $r^2 = 0.69$. In this form, the relation may be compared with those obtained by others in lakes in Japan, Canada and North America (Table 3).

Our studies agree reasonably well with the first four entries in Table 3, where the coefficient for log TP is about 1.5. Analyses of the National Eutrophication Survey (NES) data for 757 lakes by Hern et al. (1981) yields a coefficient for log TP of about 0.64 (Table 3).

While there are sound statistical reasons for the transformation of the data to logarithms, the results are not as easily interpreted. For example, any coefficient for log TP greater than 1 implies that a doubling of phosphorus in the water will result in more than a doubling of chlorophyll-a. Stated differently, the familiar crop yield-fertilizer response curve would be concave upwards rather than leveling off as more and more fertilizer is applied. If the coefficient is less than 1, as in the NES study, the yield-response curve will level and the conventional interpretation would suggest that other nutrients may be limiting. Indeed, Hern at al. (1981) indicate that 75% of the NES lakes were limited by nitrogen. However, they also suggest that light attenuation from interferences not related to chlorophyll-a can dramatically affect the quantity of algae produced per unit of total phosphorus. Lorenzen (1979) analyzed data for 493 lakes in the National Eutrophication Survey and concluded that the log CHLA-log TP relations found by most other investigators were not supported by the NES data. He proposed that phosphorus must be reduced below some threshhold level before algal production would be limited by phosphorus. This concept has been criticized by Smith and Shapiro (1981) and further discussed in exchanges of correspondence (cf. Environ. Sci. Technol. 15:1508-1511, 1981). Of the 493 lakes studied by Lorenzen (1979), 347 received municipal discharges ranging from 1 to 99% of the total phosphorus load. Thus, the lack of correlation between phosphorus and chlorophyll in such lakes is not surprising.

Returning to our data, we prefer the linear relation shown in Figure 2A. Statistical analysis of the transformed data shows that the slope of the line in Figure 2B (i.e. the coefficient for log TP) of 1.21 has a standard error of ± 0.10 and hence is not significantly different from a slope of unity. Additional analyses show that consideration of total summer nitrogen (TN) improves the linear prediction:

$$CHLA = -7.18 + 0.229\ TP + .0022\ TN$$

with multiple correlation coefficient of determination $R^2 = 0.74$. Thus, nitrogen apparently plays some role in the nutrition of algae in these lakes.

Additional evidence concerning limiting nutrients in our lakes comes from an analysis of the ratio of total N to total P in the water. Algae commonly have N/P ratios of 7/1 to 14/1 in their tissue, with a ratio of 10/1 or less in water considered to indicate nitrogen limitation (Smith and Shapiro, 1981). In only two of our 70 lakes was the N/P ratio equal to or less than 10 during the summer. In North Farms, the N/P ratio declined from 11.2 in the spring to 8.7 in the summer, and in Silver Lake the ratio declined from 19.0 to 10.2. In both lakes summer P increased dramatically because of their shallow depth, and P was not limiting. In the remaining 68, summer N/P ratios varied from 12.1 to 57.9 indicating little or no nitrogen limitation. In our previous analysis of 23 lakes (Norvell and Frink, 1975), we found that including spring nitrogen in the prediction equation for chlorophyll-a improved the coefficient of determination, but summer N improved it little.

Although Batcheldler (1981) found some evidence for the effect of Ca, Mg, and Na on algal growth in both laboratory studies and in his analysis of field data for a number of lakes, consideration of alkalinity or any of the cations and anions shown in Appendix II did not improve our prediction equation.

Transparency and Chlorophyll-a

The transparency of a lake, determined by Secchi disk, is a simple yet surprisingly good measure of the amount of algae present. The relation between transparency (S) and

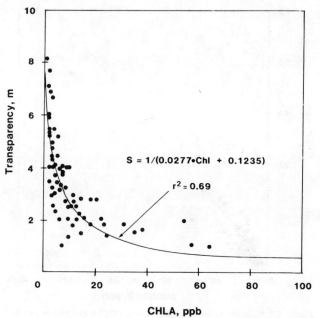

Figure 3. Relation between transparency and chlorophyll-a concentrations in 70 lakes.

chlorophyll-a is inverse as shown in Figure 3. The curve was fitted by non-linear regression analysis using the statistical program BMDPAR (Dixon et al., 1981) The equation is:

$$S = 1/(0.0277 \cdot CHLA + 0.1235)$$

with $r^2 = 0.69$. The theoretical basis for the inverse relation between transparency and light-attenuating substances is discussed in detail by Norvell and Frink (1975). Color in water also affects transparency, and a term using our measurements of color (COL) can be added to this equation. The result is:

$$S = 1/(0.0220 \cdot CHLA + 0.00341 \cdot COL + 0.099)$$

which accounts for 78% of the variability observed in transparency. The physical interpretation of the third term in the equation, i.e. $1/0.099 = 10.0$, is that turbidity and other unmeasured factors would limit the transparency to about 10 meters in the absence of chlorophyll-a and dissolved color.

Table 3. Relations between CHLA and total phosphorus (TP) in lakes.

Source	Regression Equation*	Number of Lakes	Coefficient of Determination
Jones and Bachman (1976)+	Log CHLA = –1.09 + 1.46 Log TP	143	$r^2 = 0.90$
Sakamoto (1966)†			
(as calculated by Dillon and Rigler)	Log CHLA = –1.13 + 1.58 Log TP	28	$r^2 = 0.96$
Dillon and Rigler (1974)†	Log CHLA = –1.14 + 1.45 Log TP	46	$r^2 = 0.90$
Carlson (1977)+	Log CHLA = –1.06 + 1.45 Log TP	43	$r^2 = 0.72$
Hern et al. (1981)+	Log CHLA = –0.11 + 0.64 Log TP	757	$r^2 = 0.36$

* CHLA and TP in µg/l or ppb
+ Summer CHLA and summer TP
† Summer CHLA and spring TP

Changes in Lakes During a Third of a Century

Thirty-five of the lakes were surveyed in 1937-1939 by the Lake and Pond Survey Unit of the then Connecticut State Board of Fisheries and Game (Deevey, 1940; Deevey and Bishop, 1942) permitting comparisons between conditions measured 35-40 years apart. Four of the properties were obtained by reasonably comparable methods: total P, chlorophyll, transparency, and alkalinity (Table 4). Small apparent changes should be viewed with care because of unavoidable differences in methods as well as normal day-to-day and year-to-year fluctuations in lake characteristics.

The most dramatic changes have occurred in phosphorus concentrations in the water as shown in Figure 4, where total phosphorus appears to have increased in nearly all lakes. The two lakes below the 1:1 line are Bashan (BA) and Columbia (CO), where total phosphorus appears to have decreased by 10 and 7 ppb, respectively. Bashan is a water-supply reservoir. The difference may be due to changes in lake management. Columbia Lake is drawn down in the winter to provide some weed control, which may have affected phosphorus concentrations in the water.

Accurate comparisons of chlorophyll concentrations are hindered by differences in method. Chlorophyll concentrations were probably overestimated in 1937-1939 due to interferences from turbidity and extraneous colored materials (Norvell and Frink, 1975). However, chlorophyll-a concentrations appear to have increased in most of the lakes (Figure 5). The one lake clearly below the 1:1 line is Linsley Pond (LN) where the algae and hence chlorophyll are concentrated near the thermocline, leaving the surface waters surprising clear during the summer (Norvell and Frink, 1975). A similar phenomenon occurs in Lake Wononscopomuc, but the apparent decrease in chlorophyll-a was about 1 ppb which is well within experimental error.

As expected, the transparency of most lakes has declined considerably (Figure 6). The four lakes above the 1:1 line are Silver (SL), Beseck (BS), Candlewood (CW), and Mashapaug (MH). The last two are relatively low in nutrients and their transparencies are therefore controlled in part by other factors. Both Candlewood and Beseck were noted as outliers in our earlier report. Silver Lake has experienced nearly a doubling of phosphorus and chlorophyll concentrations; hence it seems unlikely that its transparency actually increased.

Table 4. Comparison of the median, mean and range of four lake characteristics in 1937-39 and 1973-80.

Property	Year	Median	Mean	Range
Total P (ppb)	1937-1939	11.0	13.1	5.0-31
	1973-1980	16.0	25.0	7.0-106
Chlorophyll (ppb)	1937-1939	3.7	5.7	0.9-23.8
Chlorophyll-a (ppb)	1973-1980	4.5	10.0	0.8-57.0
Transparency (m)	1937-1939	3.9	4.5	1.0-8.6
	1973-1980	3.7	3.8	1.3-7.5
Alkalinity (meq/l)	1937-1939	0.26	0.57	0.09-2.43
	1973-1980	0.23	0.58	0.09-2.19

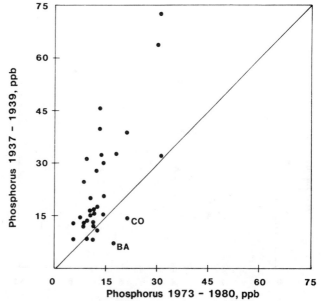

Figure 4. Total phosphorus concentrations in surface waters in 35 lakes during 1937-39 and during 1973-80.

The alkalinity of the lakes is shown in Figure 7, where it is apparent that few changes have occurred. The lake farthest below the 1:1 line is Lake Housatonic, a run-of-the-river impoundment where the alkalinity is affected by flow and waste water discharges. A number of lakes with alkalinities less than 0.5 meq/l appear to show decreases, but this is probably within experimental error. Thus, it appears that acid rain has had little impact on Connecticut lakes (Acid Rain Task Force Report, 1983).

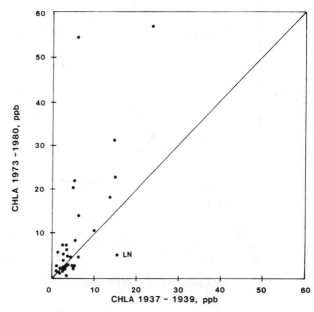

Figure 5. Chlorophyll-a concentrations in surface waters of 35 lakes during the summer in 1937-39 and in 1973-80.

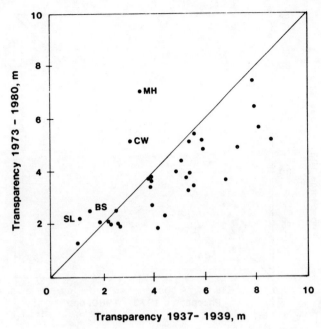

Figure 6. Transparency of 35 lakes during the summer in 1937-39 and 1973-80.

Land Use and Lake Phosphorus

Many models have been developed in the last decade to predict phosphorus loading of lakes and hence estimate their trophic status. We do not intend to review these efforts here, but will report briefly on the model developed earlier for 33 of the lakes in this study by Norvell, Frink and Hill (1979).

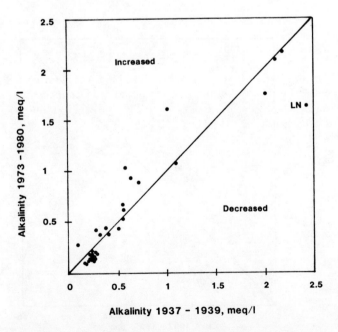

Figure 7. Alkalinity of surface waters of 35 lakes during 1937-39 and 1973-80.

Derivation of that model required a joint solution of two separate and largely independent problems. First, the regulation of phosphorus concentrations by the lake must be established. Phosphorus is removed from the water by uptake by algae and by chemical fixation in the sediments (Norvell, 1974). Lakes with long water retention times (small watershed to lake area ratios) allow for greater attenuation than do those with short retention times. Thus, we assumed that the fraction of incoming P that remains in the lake water could be related to the rate of water moving through the lake. This fraction (F) can be related to the water load on the lake surface in meters per year (Q) by the equation:

$$F = (Q + I)/(Q + V)$$

where V is the "apparent settling velocity" for P, with values of the order of 10-16 m/yr. We selected the parameter I to set a minimum value for F of 0.1 as Q approaches 0. For our 33 lake subset, F had a value of $(Q+1.2)/(Q+12)$.

The second problem is relating the total amount of P entering the lake to land use in the watershed. We assumed that the P contributed by each use was proportional to the fraction of the watershed area in that particular land use. We determined land use from a 1970 aerial survey and summarized it into three major categories: urban (U), agricultural land (A), and wooded land (W) which also indicates wetlands and the lake surface. Statistical analyses showed that the best prediction of spring P in the 33 lake subset was given by:

$$P = (Q + 1.2)/(Q + 12) \cdot (170U + 54A + 10W)/D$$

where D is the export of water from the entire watershed in meters per year. A comparison of predicted and observed phosphorus concentrations by linear regression gave a coefficient of determination of $r^2 = 0.65$ with a standard error of the estimate of phosphorus concentrations (SEM) of ± 6.9 ppb phosphorus. Estimates of phosphorus export from the three categories of land use are shown below:

Land Use	P Export Coefficients			
	kg/ha	lb/A[1]	kg/ha	lb/A[2]
Urban	1.70	1.52	1.10	0.98
Agricultural	0.54	0.48	0.91	0.81
Wooded	0.10	0.09	0.21	0.19

[1]Norvell, Frink, and Hill (1979)
[2]Median values from Reckhow et al. (1980)

In this present study, land use was obtained from a 1975 aerial survey and in somewhat different categories so that it is not directly comparable with our previous study. However, it was readily available for 63 of the lakes and serves to illustrate the utility of the model. A comparison of phosphorus predicted by our land use model and that actually observed is shown in Figure 8, where the 1:1 line is plotted.

Linear regression gave the following:

Predicted P = 7.8 + 0.74 · Observed P

with $r^2 = 0.41$. However, the equation has neither zero intercept nor unit slope, which suggests systematic differences from the earlier study. Adjustment of the terms in Q and D to account for actual runoff during the year the lake was sampled gave no improvement. Recalculation of P export coefficients decreased the estimate for urban land and increased the estimates for agricultural and wooded land closer to the median values reported by Reckhow et al. (1980). However, the improvement in the prediction of lake phosphorus was modest at best. As noted in our earlier report, corrections should be made for retention in lakes upstream. This was not done in the present study because of the large number of chain lakes requiring detailed calculations for numerous watersheds. Such corrections would be expected to improve the prediction based on our experiences with the 33-lake subset. In addition, this set of 63 lakes includes two impoundments on the Housatonic River that receive phosphorus from point sources and where observed attenuation of phosphorus is greater than that predicted by the land use model (Aylor and Frink, 1980). Omission of these two lakes from the analysis, however, does not materially alter the goodness of fit.

Despite the variability in Figure 8, the ability to estimate phosphorus in a lake based on simple estimates of land use and readily calculated lake properties seems a useful tool for watershed management (Connecticut Department of Environmental Protection, 1982). As noted by Walker (1982), the variance in prediction based on such models is likely comparable to year-to-year variability in lake water quality.

SUMMARY

A total of 70 Connecticut lakes were analyzed during 1973-1980. This report summarizes these analyses and interprets the results in terms of nutrient status and the consequent growth of weeds and algae. The data show that phosphorus is the element controlling the growth of algae in our lakes and that nitrogen is seldom limiting. The transparency of the lakes is determined largely by the growth of algae and can be predicted from measurements of chlorophyll-a and color. Comparable data for 35 lakes obtained in 1937-1939 show that total phosphorus and chlorophyll have increased in most lakes, with consequent decreases in transparency. The alkalinity of this group of lakes does not appear to have changed. Predictions of phosphorus concentrations in the lakes using a model developed earlier for a subset of 33 lakes agreed reasonably well with observed concentrations. Estimates of phosphorus export from various kinds of lands were: urban, 1.52 lb/acre; agricultural, 0.48 lb./acre; and wooded, 0.09 lb/acre. This information should help develop strategies to control nutrient enrichment so that future generations can use the land and still enjoy the lakes and ponds of Connecticut.

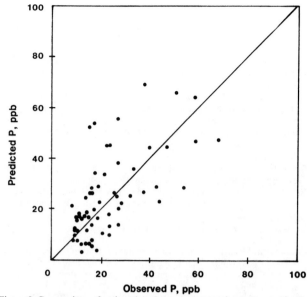

Figure 8. Comparison of spring phosphorus concentrations observed in 63 lakes with concentrations predicted by the land-use model.

ACKNOWLEDGMENTS

Samples were collected and analyzed by Jane Damschroder, David Manke, and Joseph Rapuano. The appended data and lake narratives were compiled by Jane Damschroder.

REFERENCES

Acid Rain Task Force, 1983. Acid Rain: Sources and effects in Connecticut. Report to the Governor and to the General Assembly. Conn. Agr. Exp. Sta. Bull. 809. 17p.

Aylor, D.E. and C.R. Frink. 1980. RVRFLO: A hydraulic simulator of water quality in the Housatonic River in Connecticut. Conn. Agr. Exp. Sta. Bull. 780. 53p.

Baker, L.A., P.L. Brezonik and C.R. Kratzer. 1981. Nutrient loading-trophic state relationships in Florida lakes. Florida Water Resourc. Center Publ. 56. 126p.

Bantam Lake Watershed Task Group Committee. 1981. Bantam Lake Watershed study management plan. Litchfield County Conservation District, Inc.

Batchelder, A.R. 1981. Lake trophic state and algal growth as influenced by cation concentrations. Tech. Bull. 1649. 8p.

Brooks, J.L. and E.S. Deevey, Jr. 1963. New England, p. 117-162 in G. Frey (Ed.), Limnology in North America. Univ. Wisconsin Press, Madison. 734p.

Carlson, R.E. 1977. A trophic state index for lakes. Limnol. Oceanogr. 22:361-369.

Connecticut Department of Environmental Protection. 1979a. Phase I. Diagnostic/feasibility study. Ball Pond, New Fairfield, Connecticut. Water Compliance Unit, Conn. Dept. Environ. Protection. Hartford, CT. 95p.

Connecticut Department of Environmental Protection, 1979b. Phase I. Diagnostic/feasibility study. Middle and Lower Bolton Lakes. Bolton, Vernon, Tolland, Coventry, CT. Water Compliance Unit, Conn. Dept. Environ. Protection. Hartford, CT. 116p.

Connecticut Department of Environmental Protection. 1980. Inventory of the trophic classifications of seventy Connecticut lakes. Water Compliance Unit, Conn. Dept. Environ. Protection. Hartford, CT.

Connecticut Department of Environmental Protection. 1982. A watershed management guide for Connecticut lakes. Water Compliance Unit, Conn. Dept. Environ. Protection. Hartford, CT. 21p.

Connecticut State Board of Fisheries and Game. 1959. A fishery survey of the lakes and ponds of Connecticut. Hartford, CT. 395p.

Deevey, Jr., E.S. 1940 Limnological studies in Connecticut V. A contribution to regional limnology. Am. J. Sci. 238:717-741.

Deevey, Jr., E.S. and J.S. Bishop. 1942. Section II. Limnology, p. 69-121. in State Board of Fisheries and Game, A fishery survey of important Connecticut lakes. Bull. 63. Hartford, CT. 339p.

Dillon, P.J. and F.H. Rigler. 1974. The phosphorus-chlorophyll relationship in lakes. Limnol. Oceanogr. 19:767-773.

Dixon, W.J. (Ed.) 1981. BMDP Statistical Software. Univ. of California Press. 723p.

Fairbrothers, D.E., E.T. Moul, A.R. Essbach, D.N. Riemer, and D.A. Schallock. Undated. Aquatic vegetation of New Jersey. Extension Bull. 382. Rutgers University, College of Agriculture, New Brunswick, NJ. 107p.

Frink, C.R. 1967. Nutrient budget: Rational analysis of eutrophication in a Connecticut lake. Environ. Sci. Technol. 1:425-428.

Frink, C.R. 1969. Chemical and mineralogical characteristics of eutrophic lake sediments. Soil Sci. Soc. Am. Proc. 33:326-328.

Frink, C.R. 1971. Candlewood Lake: A tentative plant nutrient budget. Conn. Agr. Exp. Sta. Circ. 238. 7p.

Frink, C.R., B.L. Sawhney, K.P. Kulp and C.G. Fredette. 1982. Polychlorinated biphenyls in Housatonic River sediments in Massachusetts and Connecticut: Determination, distribution and transport. Conn. Agr. Exp. Sta. Bull. 800, 43p.

Hern, S.C., V.W. Lambou, L.R. Williams and W.D. Taylor, 1981. Modifications of models predicting trophic state of lakes: Adjustment of models to account for the biological manifestation of nutrients. EPA-600/3-81-001. 38p.

Hotchkiss, N. 1967. Underwater and floating-leaved plants of the United States and Canada. U.S. Dept. of Interior, Bureau of Sport Fisheries and Wildlife. Resource Publ. 44. U.S. Govt. Printing Off., Washington, DC. 124p.

Hutchinson, G.E. 1957. A treatise on limnology. John Wiley & Sons, Inc. 1015p.

Jones, J.R. and R.W. Bachmann. 1976. Prediction of phosphorus and chlorophyll levels in lakes. J. Water Pollut. Control Fed. 48:2176-2182.

Jones, R.A. and G.F. Lee. 1981. Impact of phosphorus removal at the Danbury, Connecticut sewage treatment plant on water quality in Lake Lillinonah. Water, Air and Soil Pollut. 16:511-531.

Kortmann, R.W., D.D. Henry, A. Kuether, and S. Kaufman. 1982. Epilimnetic nutrient loading by metalimnetic erosion and resultant algal responses in Lake Waramaug, Connecticut. Hydrobiologia 92:501-510.

Kortmann, R.W., E. Davis, C.R. Frink and D.D. Henry. 1983. Hypolimnetic Withdrawal: Restoration of Lake Wononscopomuc, CT. p. 46-55 in: Lake Restoration, Protection and Management. Proceedings of the 2nd Annual Conference of the North American Lakes Management Society. October 26-29, 1982. Vancouver, British Columbia. EPA-44015-83-001. 327 p.

Lake Waramaug Task Force. 1978. Lake Waramaug watershed management plan. Northwestern Connecticut Regional Planning Agency, Warren, CT. 52p.

Lorenzen, M.W. 1979. Effect of phosphorus control options on lake water quality. EPA-560/11-79-011.

Muenscher, W.L. 1944. Aquatic plants of the United States. Comstock Publ. Co., Inc., Ithaca, NY. 374p.

Norvell, W.A. 1974. Insolubilization of inorganic phosphate by anoxic lake sediment. Soil Sci. Soc. Amer. Proc. 38:441-445.

Norvell, W.A. 1977. Rapid changes in the composition of Linsley and Cedar Ponds (North Branford, Connecticut). Arch. Hydrobiol. 80:286-296.

Norvell, W.A. 1980. Bantam Lake sediments: Physical and chemical properties relevant to dredging and spoils disposal. Conn. Agr. Exp. Sta. Bull. 789. 15p.

Norvell, W.A. 1982. Feasibility of inactivating phosphorus with aluminum salts in Ball Pond, CT. Conn. Agr. Exp. Sta. Bull. 806. 10p.

Norvell, W.A. and C.R. Frink. 1975. Water chemistry and fertility of twenty-three Connecticut lakes. Conn. Agr. Exp. Sta. Bull. 759. 45p.

Norvell, W.A., C.R. Frink and D.E. Hill. 1979. Phosphorus in Connecticut lakes predicted by land use. Proc. Natl. Acad. Sci. 76:5426-5429.

Reckhow, K.H., M.N. Beaulac and J.T. Simpson. 1980. Modeling phosphorus loading and lake response under uncertainty: A manual and compilation of export coefficients. EPA-440/5-80-011. 214p.

Sakamoto, M. 1966. Primary production by phytoplankton community in some Japanese lakes and its dependence on lake depth. Arch. Hydrobiol. 62:1-28.

Smith, V.H. and J. Shapiro. 1981. Chlorophyll-phosphorus relations in individual lakes. Their importance to lake restoration strategies. Environ. Sci. Technol. 15:444-451.

Thomas, M.P. 1972. Gazetteer of natural drainage areas of streams and water bodies within the state of Connecticut. Conn. Dept. Environ. Protection Bull. No. 1. 89p.

Walker, W.W. 1982. A sensitivity and error analysis framework for lake eutrophication modeling. Water Res. Bull. 18:53-60.

APPENDIX I

Locations and names of lakes

Map Location	Name of Lake	Town Location
1	Alexander Lake	Killingly
2	Amos Lake	Preston
3	Ball Pond	New Fairfield
4	Bantam Lake	Litchfield, Morris
5	Bashan Lake	East Haddam
6	Batterson Park Pond	Farmington, New Britain
7	Beach Pond	Voluntown
8	Beseck Lake	Middlefield
9	Bigelow Pond	Union
10	Billings Lake	North Stonington
11	Black Pond	Meriden, Middlefield
12	Black Pond	Woodstock
13	Burr Pond	Torrington
14	Candlewood Lake	New Fairfield, Sherman, Danbury New Milford, Brookfield
15	Cedar Pond	North Branford
16	Cedar Lake	Chester
17	Columbia Lake	Columbia
18	Cream Hill Pond	Cornwall
19	Crystal Lake	Ellington, Stafford
20	Dodge Pond	East Lyme
21	Eagleville Lake	Mansfield
22	East Twin Lake	Salisbury
23	1860 Reservoir	Wethersfield
24	Gardner Lake	Salem, Montville, Bozrah
25	Glasgo Pond	Griswold
26	Gorton Pond	East Lyme
27	Hayward Lake	East Haddam
28	Highland Lake	Winchester
29	Hitchcock Lake	Wolcott
30	Housatonic Lake	Shelton
31	Kenosia Lake	Danbury
32	Lillinonah Lake	Southbury, Bridgewater, Newtown, Brookfield
33	Linsley Pond	North Branford, Branford
34	Little Pond	Thompson
35	Long Pond	Ledyard, North Stonington
36	Long Meadow Pond	Bethlehem
37	Lower Bolton Lake	Bolton, Vernon
38	Mamanasco Lake	Ridgefield
39	Mashapaug Lake	Union
40	Middle Bolton Lake	Bolton
41	Moodus Reservoir	East Haddam
42	Mount Tom Pond	Litchfield, Morris, Washington
43	Mudge Pond	Sharon
44	North Farms Reservoir	Wallingford

APPENDIX I
(continued)

Map Location	Name of Lake	Town Location
45	Norwich Pond	Lyme
46	Pachaug Pond	Griswold
47	Pataganset Lake	East Lyme
48	Pocotopaug Lake	East Hampton
49	Powers Lake	East Lyme
50	Quaddick Reservoir	Thompson
51	Quassapaug Lake	Middlebury
52	Quonnipaug Lake	Guilford
53	Rogers Lake	Lyme, Old Lyme
54	Roseland Lake	Woodstock
55	Shenipsit Lake	Vernon, Ellington, Tolland
56	Silver Lake	Berlin, Meriden
57	Squantz Pond	New Fairfield, Sherman
58	Taunton Pond	Newtown
59	Terramuggus Lake	Marlborough
60	Tyler Pond	Goshen
61	Uncas Lake	Lyme
62	Waramaug Lake	Warren, Washington, Kent
63	Waumgumbaug Lake	Coventry
64	West Hill Pond	New Hartford
65	West Side Pond	Goshen
66	Winnemaug Lake	Watertown
67	Wononpakook Lake	Salisbury
68	Wononscopomuc Lake	Salisbury
69	Wyassup Lake	North Stonington
70	Zoar Lake	Newtown, Monroe, Oxford, Southbury

APPENDIX II

Selected physical and chemical properties of 70 lakes.

Lake	Surface area[a]	Watershed[a,b] Lake area	Mean[a] depth	Max.[a] depth	Year Sampled	Summer transparency	Summer color	Alkalinity[c]	Calcium[c]	Magnesium[c]	Sodium[c]	Potassium[c]	Chloride[c]
	ha		---- m ----			m	ppm	------------------- meq/l ---------------------					
Alexander	76.1	4.0	7.4	16.2	1974	7.2	7	0.12	0.16	0.05	0.13	0.03	0.06
Amos	42.0	5.1	5.8	14.5	1980	3.7	5	0.38	0.44	0.18	0.27	0.06	0.26
Ball	36.0	2.6	6.9	15.8	1980	2.5	5	0.98	0.98	0.43	0.40	0.05	0.74
Bantam	366.4	23.2	4.4	7.6	1974	1.8	20	0.61	0.45	0.32	0.27	0.03	0.20
Bashan	110.5	4.6	4.8	14.5	1980	5.5	5	0.08	0.27	0.11	0.22	0.02	0.12
Batterson Park	65.1	16.8	4.5	6.1	1979	1.5	45	0.66	0.84	0.39	0.44	0.02	1.11
Beach	157.7	7.6	6.1	19.7	1979	7.2	12	0.04	0.15	0.06	0.22	0.01	0.16
Beseck	47.8	11.2	3.5	7.3	1974	2.8	10	0.53	0.51	0.29	0.32	0.02	0.20
Bigelow	7.4	47.0	2.3	4.8	1980	2.5	40	0.10	0.21	0.12	0.27	0.02	0.20
Billings	42.0	4.3	4.2	10.0	1980	4.5	20	0.10	0.22	0.07	0.18	0.02	0.08
Black (Meriden)	30.2	10.0	2.6	7.0	1979	2.5	17	0.72	0.76	0.38	0.39	0.02	0.63
Black (Woodstock)	29.4	6.3	3.7	7.0	1980	3.0	20	0.14	0.19	0.11	0.23	0.02	0.16
Burr	34.0	9.9	1.5	3.9	1980	2.7	15	0.10	0.29	0.11	0.32	0.03	0.45
Candlewood[d]	2195.0	8.1	8.9	25.9	1974	5.2	5	0.94	0.72	0.41	0.28	0.03	0.20
Cedar (No. Branford)[e]	8.7	25.2	3.3	5.2	1974	0.9	15	1.76	0.80	0.48	1.27	0.03	0.23
Cedar (Chester)	27.2	30.4	5.8	13.6	1979	4.1	20	0.10	0.18	0.09	0.18	0.02	0.08
Columbia	112.7	6.9	5.1	7.8	1980	5.0	10	0.22	0.23	0.09	0.24	0.03	0.12
Cream Hill	28.8	5.6	4.8	13.0	1978	5.5	10	0.92	0.45	0.22	0.22	0.02	
Crystal	80.4	8.9	5.9	15.2	1979	4.0	8	0.14	0.23	0.10	0.28	0.02	0.27
Dodge	13.2	13.0	6.1	14.5	1979	4.0	15	0.15	0.28	0.13	0.27	0.03	0.26
Eagleville	32.0	888.0	0.9	3.0	1980	1.5	>70	0.18	0.25	0.12	0.30	0.03	0.18
East Twin	224.9	4.7	9.9	24.4	1974	5.3	9	2.11	1.43	0.88	0.12	0.04	0.08
1860 Reservoir	14.2	14.8	0.6	1.1	1979	1.0	40	1.43	1.23	0.68	0.32	0.03	0.40
Gardner	194.7	7.3	4.2	13.1	1974	4.1	15	0.20	0.18	0.07	0.27	0.02	0.17
Glasgo	73.7	131.3	3.0	7.6	1980	2.0	45	0.09	0.18	0.08	0.20	0.03	0.09
Gorton	21.5	78.4	1.1	2.3	1979	2.0	25	0.19	0.30	0.15	0.28	0.04	0.35
Hayward	79.6	8.0	3.1	11.3	1974	3.3	15	0.15	0.12	0.08	0.19	0.03	0.06
Highland	177.6	10.2	6.0	18.8	1979	6.0	10	0.21	0.30	0.14	0.35	0.03	0.40
Hitchcock	47.4	3.0	1.9	3.0	1979	2.0	25	0.40	0.51	0.17	0.47	0.03	0.71
Housatonic	131.3	3068.9	2.8	7.9	1980	2.0	10	1.36	1.06	0.66	0.34	0.04	0.26
Kenosia	22.7	56.2	3.5	4.6	1980	1.8	5	0.88	0.82	0.40	0.29	0.04	0.27
Lillinonah	769.5	470.0	11.8	30.5	1974	1.6	10	1.54	1.21	0.71	0.39	0.04	0.23
Linsley[e]	9.3	35.5	6.3	13.4	1974	3.5	15	1.61	0.90	0.51	0.95	0.03	0.28
Little	26.2	7.0	2.4	4.2	1980	3.0	20	0.23	0.37	0.13	0.24	0.04	0.22
Long	39.9	29.6	4.6	22.0	1974	4.8	10	0.19	0.19	0.10	0.22	0.02	0.14

APPENDIX II
(continued)

Lake	Surface area[a]	Watershed[a,b] Lake area	Mean[a] depth	Max.[a] depth	Year Sampled	Summer transparency	Summer color	Alkalinity[c]	Calcium[c]	Magnesium[c]	Sodium[c]	Potassium[c]	Chloride[c]
	ha		---- m ----			m	ppm			------ meq/l ------			
Long Meadow	44.2	9.4	1.3	2.1	1980	1.3	5	0.18	0.33	0.14	0.22	0.04	0.13
Lower Bolton	71.4	13.6	3.4	7.9	1979	2.3	25	0.12	0.26	0.10	0.27	0.03	0.26
Mamanasco	38.0	5.7	2.1	3.0	1979	1.8	20	1.10	1.27	0.50	0.40	0.05	0.83
Mashapaug	118.8	10.2	2.8	13.0	1979	8.2	5	0.09	0.22	0.13	0.27	0.02	0.35
Middle Bolton	46.0	16.9	3.0	7.9	1979	2.5	50	0.09	0.24	0.11	0.24	0.03	0.26
Moodus	180.4	14.9		3.0	1980	2.0	23	0.10	0.21	0.10	0.23	0.04	0.08
Mt. Tom	24.6	11.2	6.4	13.9	1979	4.5	20	0.38	0.42	0.20	0.28	0.04	0.33
Mudge	80.4	36.6	6.7	10.7	1974	4.1	7	2.63	1.43	1.38	0.18	0.04	0.14
North Farms	25.0	7.6	0.9	1.5	1979	1.0	30	0.57	0.71	0.21	0.23	0.08	0.20
Norwich	11.0	11.2	6.9	11.5	1980	3.0	50	0.28	0.29	0.14	0.19	0.02	0.05
Pachaug	332.4	40.2	1.8	5.5	1979	3.5	60	0.11	0.22	0.09	0.18	0.02	0.15
Pataganset	49.2	20.3	3.8	10.7	1974	2.8	25	0.19	0.11	0.09	0.24	0.02	0.11
Pocotopaug	204.7	5.6	3.5	11.6	1974	3.6	11	0.14	0.19	0.09	0.27	0.03	0.20
Powers	61.0	4.0	2.1	3.9	1980	3.2	10	0.11	0.18	0.08	0.27	0.02	0.06
Quaddick	186.7	33.5	1.9	7.6	1980	2.3	50	0.10	0.22	0.08	0.20	0.02	0.06
Quassapaug	108.4	4.3	8.7	19.8	1974	6.8	3	0.13	0.17	0.11	0.18	0.03	0.17
Quonnipaug	44.6	15.1	4.1	14.5	1980	4.4	15	0.74	0.47		0.23	0.02	0.14
Rogers	106.0	18.2	6.1	20.0	1978	4.0	20	0.22	0.17	0.08			0.17
Roseland	35.2	221.0	3.1	6.1	1974	2.8	20	0.45	0.47	0.14	0.22	0.06	0.17
Shenipsit	209.1	20.1	9.2	20.7	1974	4.0	15	0.16	0.16	0.10	0.23	0.03	0.14
Silver	60.4	8.4	1.4	3.6	1979	1.8	25	0.73	1.00	0.49	0.48	0.02	0.84
Squantz	115.2	12.6	6.9	14.2	1980	3.4	5	0.34	0.44	0.03	0.20	0.03	0.17
Taunton	50.6	6.7	6.6	9.0	1974	3.3	10	0.45	0.44	0.23	0.37	0.03	0.34
Terramuggus	33.2	4.1	6.5	13.1	1974	6.0	7	0.23	0.29	0.16	0.48	0.04	0.35
Tyler	72.8	22.2	3.7	7.9	1979	3.8	25	0.60	0.48	0.25	0.17	0.03	0.09
Uncas	27.6	13.9	6.7	12.1	1980	5.4	5	0.18	0.23	0.11	0.20	0.02	0.09
Waramaug	272.1	13.4	6.7	12.2	1974	2.7	10	0.39	0.31	0.19	0.25	0.03	0.14
Waumgumbaug	151.0	5.6	8.8	12.1	1980	6.1	5	0.31	0.41	0.12	0.34	0.05	0.32
West Hill	95.3	3.5	9.7	18.0	1974	7.0	5	0.14	0.15	0.07	0.12	0.02	0.06
West Side	17.0	51.4	4.6	10.0	1978	4.0	25	0.75	0.80	0.36	0.16	0.02	0.09
Winnemaug	48.0	5.7	2.4	5.0	1980	1.3	15	0.38	0.39	0.23	0.29	0.07	0.30
Wononpakook	65.6	16.9	3.5	7.4	1980	2.1	20	2.86	1.90	1.50	0.23	0.07	0.27
Wononscopomuc	141.0	4.7	11.1	32.9	1974	7.8	5	2.19	1.12	1.23	0.25	0.04	0.23
Wyassup	37.1	5.6	2.7	8.5	1979	4.3	17	0.10	0.20	0.09	0.20	0.02	0.09
Zoar	394.8	1014.0	7.5	22.9	1974	1.9	12	1.76	1.14	0.68	0.38	0.04	0.28

[a](Connecticut) State Board of Fisheries and Game, 1959.
[b]Thomas, M.D., 1972.
[c]Values for 1974 are mean concentrations for spring, summer, and fall.
 Values for 1978, 1979, and 1980 are concentrations for spring overturn.
[d]Includes pump-up from the Housatonic River.
[e]Includes recent expansion of quarrying into the watershed.

APPENDIX III

All analyses for nutrients, transparency and alkalinity, 1973-1980, where *comp.* indicates a composite sample, and *nd* means not determined.

Lake and date sampled	Trans., m	Depth, m	Alk., meq/l	Chl-a	Sol. P	Total P	NH$_4$-N	NO$_3$-N	Sol. N	Total N
ALEXANDER							---ppb---			
102373	5.7	0.2	0.10	nd	4	8	40	10	200	250
		5.0	nd	nd	3	8	40	10	190	250
		10.0	nd	nd	3	8	50	10	190	250
		13.0	nd	nd	4	16	190	20	240	370
50274	9.7	0.2	0.12	nd	2	7	40	40	280	330
		4.0	nd	nd	5	10	40	40	240	260
		8.0	nd	nd	5	13	30	40	260	300
		13.0	nd	nd	5	14	10	40	240	320
71774	6.3	0-3	0.13	0.6	3	6	0	40	310	340
		4.0	nd	nd	3	8	0	50	320	360
		8.0	nd	nd	3	14	0	50	290	370
		13.0	nd	nd	5	14	140	110	510	600
82774	8.2	0-3	0.11	0.9	3	9	60	70	380	470
		5.0	nd	nd	3	8	40	30	260	370
		9.0	nd	nd	4	13	40	30	280	410
		13.0	nd	nd	nd	16	30	60	nd	380
AMOS										
40380	2.6	comp	0.38	nd	5	23	7	235	395	516
81880	3.7	0-4	0.42	4.0	6	16	25	9	229	339
		6.0	0.44	nd	7	11	25	0	235	385
		8.0	0.46	nd	9	46	30	1	206	501
		11.0	0.56	nd	12	33	643	0	810	1125
BALL										
41180	1.3	comp	0.98	nd	14	37	45	193	598	888
72480	2.5	0-4	1.08	3.0	7	23	26	4	424	544
		5.0	1.04	nd	8	40	10	0	327	627
		7.0	1.08	nd	15	42	206	10	383	677
		9.0	1.30	nd	34	66	900	0	1053	1373
		11.0	1.40	nd	156	198	1694	8	2048	2408
		13.0	1.43	nd	237	252	2088	0	2440	2760
		14.0	1.82	nd	246	345	2200	0	2520	2853
BANTAM										
92673	2.0	0.2	nd	nd	18	36	90	60	nd	670
		5.0	nd	nd	16	35	110	50	nd	680
42474	3.0	0.2	0.50	nd	9	25	20	160	360	520
		5.0	nd	nd	8	29	0	50	290	440
62474	1.8	0.2	0.58	50.0	7	28	50	30	300	660
		3.0	nd	nd	7	28	20	40	280	700
		5.0	nd	nd	nd	40	80	100	nd	710
72374	1.5	0-2	0.70	30.0	7	37	40	60	520	1150
82274	2.2	0-3	0.67	14.0	31	40	80	10	670	870
		5.0	nd	nd	10	70	100	10	550	940
BASHAN										
40780	6.0	comp	0.08	nd	5	8	4	30	115	205
72280	5.5	0-6	0.10	2.0	2	5	24	8	158	193
		8.0	0.12	nd	6	11	28	10	105	180
		10.0	0.14	nd	16	16	28	14	129	229
		12.0	0.20	nd	15	15	169	23	278	333
BATTERSON PARK										
42479	2.3	comp	0.66	nd	9	40	nd	48	328	508
82979	1.5	0-3	0.81	35.0	6	31	25	8	446	933
		5.0	nd	nd	12	37	104	18	518	787

APPENDIX III (continued)

Lake and date sampled	Trans., m	Depth, m	Alk., meq/l	Chl-a	Sol. P	Total P	NH_4-N	NO_3-N	Sol. N	Total N
						---ppb---				
BEACH										
40479	6.0	comp	0.04	nd	3	12	40	34	244	244
80379	7.2	0-5	0.09	1.0	2	5	22	0	163	200
		7.0	nd	nd	3	6	4	11	186	249
		10.0	nd	nd	3	7	5	11	86	149
		15.0	nd	nd	3	8	16	60	260	298
		19.0	nd	nd	3	5	40	96	334	371
BESECK										
103173	2.0	0.2	0.59	nd	7	36	110	70	360	780
		3.0	nd	nd	11	38	100	90	600	800
		6.0	nd	nd	10	36	120	60	420	750
40374	2.2	0.2	0.34	nd	10	21	0	260	410	530
		3.0	nd	nd	9	27	10	340	480	600
		5.0	nd	nd	8	34	10	340	480	540
62174	4.0	0.2	0.54	7.2	6	15	20	60	310	410
		4.0	nd	nd	7	22	30	20	220	340
		6.0	nd	nd	nd	35	180	40	nd	650
72374	2.5	0-3	0.55	13.9	7	34	50	50	250	570
82674	2.0	0-3	0.64	34.0	10	52	50	30	430	960
		5.0	nd	nd	nd	43	160	150	nd	850
		6.0	nd	nd	nd	165	950	150	nd	1580
BIGELOW										
41880	5.2	comp	0.10	nd	9	9	45	10	95	155
81980	2.5	0-3	0.14	9.0	4	11	35	1	221	281
BILLINGS										
40380	5.2	comp	0.10	nd	4	8	5	9	69	184
82080	4.5	0-4	0.10	3.0	2	8	25	0	244	320
		6.0	0.13	nd	3	11	25	0	150	240
		8.0	0.14	nd	5	14	16	0	140	235
BLACK (Meriden)										
32679	2.3	comp	0.72	nd	5	25	30	0	340	500
83079	2.5	0-3	0.87	10.0	4	17	25	0	375	550
		4.0	nd	39.0	15	70	20	0	338	1175
		6.0	nd	nd	8	70	20	0	301	925
BLACK (Woodstock)										
41880	5.2	comp	0.14	nd	10	12	10	4	159	239
81980	3.0	0-3	0.18	11.0	4	15	30	0	270	500
		5.0	0.18	nd	4	29	46	0	235	460
		6.0	0.42	nd	4	23	57	1	221	376
BURR										
42180	3.7	comp	0.10	nd	7	11	15	56	186	251
80880	2.7	0-3	0.14	8.0	6	17	20	18	291	498
CANDLEWOOD										
112073	5.4	0.2	0.97	nd	8	19	60	90	320	410
		5.0	nd	nd	9	24	50	90	310	370
		10.0	nd	nd	16	24	80	100	280	330
		15.0	nd	nd	34	61	130	340	340	660
50874	5.3	0.2	0.92	nd	3	10	10	60	220	330
		5.0	nd	nd	3	14	10	60	260	370
		10.0	nd	nd	3	14	10	60	240	390
		20.0	nd	nd	4	17	40	60	240	370
71574	5.7	0-3	0.80	2.4	6	15	40	20	330	440
		5.0	nd	nd	6	17	40	50	230	420
		10.0	nd	nd	6	20	40	70	250	330
		20.0	nd	nd	nd	27	190	510	nd	770
81374	4.5	0-3	0.91	7.5	4	16	40	0	320	520
91074	5.3	0-3	0.97	3.6	3	13	0	30	280	350
		5.0	nd	nd	4	17	10	30	220	310
		10.0	nd	nd	4	38	150	30	400	480
		15.0	nd	nd	14	49	80	200	nd	420
		20.0	nd	nd	15	47	330	160	nd	680

APPENDIX III (continued)

Lake and date sampled	Trans., m	Depth, m	Alk., meq/l	Chl-a	Sol. P	Total P	NH₄-N	NO₃-N	Sol. N	Total N
						---ppb---				
CEDAR POND										
110973	2.0	0.2	2.20	nd	24	80	380	100	970	1450
		2.0	nd	nd	22	89	350	70	960	1470
		4.0	nd	nd	24	94	360	80	880	1470
40374	1.1	0.2	1.50	nd	8	42	30	1050	1380	1520
		2.0	nd	nd	9	48	30	1050	1330	1580
		4.0	nd	nd	9	55	60	1060	1340	1550
72374	0.9	0-2	1.70	64.0	16	71	20	40	430	1830
		4.0	nd	nd	16	98	80	10	430	1470
CEDAR LAKE										
32779	4.3	comp	0.10	nd	5	14	2	88	389	413
81579	4.1	0-5	0.13	9.0	5	17	20	10	210	285
		6.0	nd	nd	6	14	16	0	125	238
		8.0	nd	nd	5	13	15	10	135	248
		10.0	nd	nd	12	16	92	0	275	538
COLUMBIA										
40880	3.0	comp	0.22	nd	10	18	0	49	124	259
80580	5.0	0-4	0.13	3.0	5	10	16	13	238	263
		6.0	0.10	nd	5	13	16	20	255	305
CREAM HILL										
51178	nd	comp	0.92	nd	6	16	250	0	150	400
72878	5.5	0-5	0.97	4.0	10	18	nd	0	400	530
		7.0	nd	nd	10	23	nd	0	280	330
		9.0	nd	nd	22	36	nd	0	350	530
CRYSTAL										
41079	3.5	comp	0.14	nd	3	12	20	48	328	373
81379	4.0	0-6	0.26	8.0	2	12	22	0	225	363
		7.0	nd	nd	1	8	0	0	188	250
		10.0	nd	nd	3	8	8	18	218	293
		14.0	nd	nd	5	18	112	8	233	296
DODGE										
43079	4.8	comp	0.15	nd	6	15	15	204	654	779
80779	4.0	0-4	0.34	2.0	5	12	8	0	275	400
		5.0	nd	nd	3	12	2	0	238	338
		9.0	nd	nd	3	13	40	187	425	487
		13.0	nd	nd	7	9	605	18	nd	868
EAGLEVILLE										
50980	2.3	comp	0.18	nd	15	34	100	198	433	533
80680	1.5	0-1.5	0.36	14.0	31	60	55	194	559	729
EAST TWIN										
101773	6.2	0.2	2.03	nd	2	10	20	70	380	410
		5.0	nd	nd	3	12	20	30	240	310
		10.0	nd	nd	2	8	20	20	280	410
		15.0	nd	nd	7	17	110	10	390	560
		20.0	nd	nd	52	72	640	0	790	910
50774	5.3	0.2	2.32	nd	9	21	40	40	250	470
		5.0	nd	nd	9	22	10	40	320	450
		10.0	nd	nd	10	17	40	40	320	400
		15.0	nd	nd	9	19	40	40	290	380
		20.0	nd	nd	12	21	40	20	330	380
71074	5.0	0-3	2.12	0.6	2	14	30	80	360	530
		5.0	nd	nd	5	11	20	80	320	430
		10.0	nd	nd	5	13	20	80	310	340
		15.0	nd	nd	4	13	60	130	440	490
		20.0	nd	nd	nd	46	nd	nd	nd	520
73074	6.0	0-3	2.11	1.5	8	17	0	0	240	420
90474	5.0	0-3	1.96	4.8	4	16	20	20	410	580
		5.0	nd	nd	4	13	0	20	230	260
		10.0	nd	nd	5	15	20	20	240	380
		15.0	nd	nd	7	25	40	80	270	380
		20.0	nd	nd	134	141	560	20	810	950

APPENDIX III (continued)

Lake and date sampled	Trans., m	Depth, m	Alk., meq/l	Chl-a	Sol. P	Total P	NH$_4$-N	NO$_3$-N	Sol. N	Total N
							---ppb---			
1860 RESERVOIR										
42479	1.0	comp	1.43	nd	21	59	45	18	635	1158
100179	1.0	0-1	1.21	7.0	2	32	144	20	1058	1260
GARDNER										
111373	3.3	0.2	0.17	nd	9	28	40	60	430	510
		4.0	nd	nd	8	30	50	100	330	480
		7.0	nd	nd	9	34	30	50	260	410
		10.0	nd	nd	7	28	20	80	260	390
42574	3.5	0.2	0.12	nd	5	13	10	210	330	450
		3.0	nd	nd	4	16	10	230	370	500
		6.0	nd	nd	4	15	0	230	380	450
		9.0	nd	nd	6	14	0	230	390	510
70174	3.5	0.2	0.24	9.0	4	14	0	30	330	450
		3.0	nd	nd	5	15	0	0	270	400
		6.0	nd	nd	4	18	0	0	350	510
		9.0	nd	nd	nd	22	20	60	nd	780
82174	4.8	0-3	0.26	4.8	5	12	20	80	430	740
		6.0	nd	nd	5	15	20	40	360	470
		10.0	nd	nd	nd	24	420	60	nd	970
GLASGO										
40980	2.8	comp	0.09	nd	9	13	21	177	322	357
82280	2.0	0-3	0.18	9.0	4	19	13	6	236	396
		5.0	0.70	nd	8	21	193	10	455	645
GORTON										
32779	2.5	comp	0.19	nd	6	17	68	396	456	786
80779	2.0	0-2	0.51	6.0	7	19	20	0	300	475
HAYWARD										
112073	3.3	0.2	0.19	nd	5	13	20	90	390	510
		3.0	nd	nd	7	16	10	80	350	460
		6.0	nd	nd	8	21	10	60	330	590
42574	4.8	0.2	0.10	nd	5	24	50	150	290	420
		3.0	nd	nd	1	21	30	170	320	380
		6.0	nd	nd	1	22	30	70	210	260
72974	3.3	0-2	0.17	7.8	5	15	50	10	220	370
		3.0	nd	nd	3	14	10	10	130	200
		6.0	nd	nd	5	16	40	0	110	190
HIGHLAND										
41179	4.3	comp	0.21	nd	3	11	20	80	200	280
80279	6.0	0-5	0.67	2.0	0	5	25	0	100	238
		6.0	nd	nd	2	6	20	0	163	200
		10.0	nd	nd	1	19	0	25	125	263
		15.0	nd	nd	1	11	22	204	304	404
		18.0	nd	nd	1	18	748	56	469	581
HITCHCOCK										
50779	3.0	comp	0.40	nd	9	17	14	0	280	580
82879	2.0	0-2	0.53	12.0	6	22	12	0	313	538
HOUSATONIC										
50580	2.1	comp	1.36	nd	15	41	50	412	602	747
82880	2.0	0-8	1.90	22.0	10	51	100	105	475	700
KENOSIA										
41580	1.0	comp	0.88	nd	11	38	27	400	575	770
73180	1.8	0-1	1.48	17.0	8	20	12	18	303	508
		3.0	1.50	nd	7	45	20	6	286	741
		4.5	1.68	nd	15	62	28	9	339	834

APPENDIX III (continued)

Lake and date sampled	Trans., m	Depth, m	Alk., meq/l	Chl-a	Sol. P	Total P	NH₄-N	NO₃-N	Sol. N	Total N
							ppb			
LILLINONAH										
101973	2.0	0.2	1.95	nd	19	48	90	250	540	790
		5.0	nd	nd	27	56	120	300	640	790
		10.0	nd	nd	19	59	110	280	620	820
		15.0	nd	nd	23	57	130	280	590	770
		20.0	nd	nd	41	62	320	190	720	770
		25.0	nd	nd	nd	119	340	270	nd	910
53174	3.2	0.2	1.18	nd	16	45	50	230	480	720
		5.0	nd	nd	23	55	70	290	420	730
		10.0	nd	nd	25	68	130	290	530	730
		15.0	nd	nd	nd	62	170	310	nd	760
		20.0	nd	nd	nd	77	140	340	nd	820
		23.0	nd	nd	nd	76	340	340	nd	870
70874	1.3	0.2	nd	9.3	29	75	40	150	640	1180
		5.0	nd	nd	37	69	100	260	750	950
		10.0	nd	nd	34	63	180	260	750	890
		15.0	nd	nd	38	48	220	220	870	970
		20.0	nd	nd	nd	66	240	420	nd	1120
		25.0	nd	nd	nd	22	110	260	nd	830
72574	2.5	0-3	1.23	7.8	11	35	60	110	480	610
91274	1.9	0-3	1.78	38.0	13	59	20	120	590	1010
		5.0	nd	nd	28	54	50	240	530	710
		10.0	nd	nd	38	83	90	220	590	710
		15.0	nd	nd	50	61	90	250	600	670
		20.0	nd	nd	59	68	110	250	640	640
		25.0	nd	nd	53	78	140	300	720	770
LINSLEY										
110973	1.5	0.2	1.83	nd	9	36	260	0	540	710
		3.0	nd	nd	7	29	250	0	450	680
		7.0	nd	nd	8	38	280	120	640	830
		11.0	nd	nd	nd	177	2100	150	2300	2750
40374	1.1	0.2	1.30	nd	8	54	30	750	1060	1340
		5.0	nd	nd	8	48	30	680	990	1180
		10.0	nd	nd	10	47	0	880	1280	1440
80974	3.5	0-3	1.71	5.2	8	28	30	80	370	520
		5.0	nd	nd	8	37	20	20	250	610
		9.0	nd	nd	nd	122	970	100	nd	1430
LITTLE										
41780	3.1	comp	0.23	nd	19	13	33	379	629	644
82580	3.0	0-3	0.32	4.0	14	23	28	10	275	365
LONG										
112073	3.2	0.2	0.21	nd	5	14	20	60	260	330
		5.0	nd	nd	6	16	10	40	140	250
		10.0	nd	nd	6	14	10	160	280	350
		15.0	nd	nd	4	14	40	80	280	400
		19.0	nd	nd	9	14	40	30	170	200
41074	3.2	0.2	0.15	nd	4	11	10	160	280	300
		5.0	nd	nd	3	13	30	220	310	360
		10.0	nd	nd	3	11	10	150	220	270
		15.0	nd	nd	5	8	30	160	260	300
		19.0	nd	nd	8	9	30	140	190	260
80674	4.8	0-3	0.20	2.8	4	13	40	110	420	640
		5.0	nd	nd	5	11	10	80	200	330
		10.0	nd	nd	3	9	10	90	180	250
		15.0	nd	nd	3	9	30	180	200	300
		19.0	nd	nd	nd	10	40	170	nd	300

APPENDIX III (continued)

Lake and date sampled	Trans., m	Depth, m	Alk., meq/l	Chl-a	Sol. P	Total P	NH$_4$-N	NO$_3$-N	Sol. N	Total N
							----ppb----			
LONG MEADOW										
41480	1.1	comp	0.18	nd	12	38	5	251	391	586
72880	1.3	0-2	0.42	9.0	4	29	33	11	358	611
LOWER BOLTON										
41779	3.3	comp	0.12	nd	14	19	8	48	218	358
90179	2.3	0-3	0.17	13.0	4	15	75	19	682	869
		4.0	nd	nd	3	17	112	19	336	569
		6.0	nd	nd	27	52	264	26	614	651
MAMANASCO										
42079	3.0	comp	1.10	nd	7	27	30	18	358	598
82979	1.8	0-2	1.40	11.0	9	34	20	0	363	575
MASHAPAUG										
41079	6.0	comp	0.09	nd	4	9	8	42	187	193
81379	8.2	0-8	0.11	1.0	1	6	22	0	225	325
		9.0	nd	nd	1	5	16	0	138	188
		12.0	nd	nd	2	9	12	0	140	152
MIDDLE BOLTON										
42479	2.5	comp	0.09	nd	6	28	2	0	310	470
82179	2.5	0-4	0.16	12.0	15	24	125	19	507	757
		5.0	nd	nd	12	33	347	0	650	913
MOODUS										
40780	2.1	comp	0.10	nd	22	33	14	0	123	345
72580	2.0	0-1	0.22	9.0	9	22	33	12	324	519
		2.0	0.22	nd	5	24	44	12	347	504
MOUNT TOM										
40579	3.5	comp	0.38	nd	10	10	30	66	406	456
82079	4.5	0-5	0.38	5.0	2	11	25	0	317	375
		6.0	nd	13.0	8	20	40	20	345	420
		8.0	nd	nd	10	38	92	0	275	463
		10.0	nd	nd	63	141	625	0	850	1075
		13.0	nd	nd	215	258	885	0	1100	1400
MUDGE										
110773	2.5	0.2	3.11	nd	11	33	210	60	590	720
		4.0	nd	nd	12	35	220	50	450	720
		8.0	nd	nd	11	30	200	60	570	680
43074	2.5	0.2	3.02	nd	1	23	40	140	330	520
		4.0	nd	nd	1	27	40	150	330	550
		8.0	nd	nd	1	33	40	210	440	620
71174	4.5	0-3	2.66	2.0	5	17	20	40	330	460
		4.0	nd	nd	1	12	20	80	320	390
		9.0	nd	nd	nd	32	180	60	nd	500
73074	3.8	0-3	2.62	4.2	5	17	0	40	160	400
82874	4.0	0-3	1.74	5.6	10	22	50	10	430	590
		5.0	nd	nd	7	18	50	0	260	480
		9.0	nd	nd	nd	64	150	60	nd	700
NORTH FARMS										
32679	1.5	comp	0.57	nd	23	54	43	0	550	610
90179	1.0	0-1	1.20	57.0	84	157	54	0	775	1375
NORWICH										
40280	3.2	comp	0.28	nd	10	15	12	17	162	217
73080	3.0	0-2	0.19	3.0	6	12	20	0	110	295
		3.0	0.20	nd	6	13	25	0	230	315
		5.0	0.24	nd	8	11	33	8	208	223
		7.0	0.38	nd	15	15	29	30	220	240
		9.0	0.44	nd	12	15	50	82	272	282
PACHAUG										
50279	3.0	comp	0.11	nd	7	15	22	143	676	nd
81479	3.5	0-3	0.29	2.0	4	16	28	20	383	420
		4.0	nd	nd	6	13	32	20	345	345

APPENDIX III (continued)

Lake and date sampled	Trans., m	Depth, m	Alk., meq/l	Chl-a	Sol. P	Total P	NH$_4$-N	NO$_3$-N	Sol. N	Total N
							---ppb---			
PATAGANSET										
111373	3.0	0.2	0.17	nd	4	13	80	50	300	370
		5.0	nd	nd	4	11	60	50	310	390
		9.0	nd	nd	3	14	80	50	280	380
41074	3.0	0.2	0.06	nd	5	15	10	180	270	460
		5.0	nd	nd	8	14	30	100	190	320
		9.0	nd	nd	8	16	40	90	180	310
70174	2.5	0.2	0.34	14.3	3	15	0	40	400	760
		5.0	nd	nd	3	12	0	40	360	470
		9.0	nd	nd	nd	15	190	150	nd	860
82174	3.0	0-3	0.19	13.8	3	15	0	40	390	600
		5.0	nd	nd	3	17	0	40	340	570
		9.0	nd	nd	nd	15	30	270	nd	830
POCOTOPAUG										
111573	4.5	0.2	0.12	nd	8	16	70	50	310	320
		4.0	nd	nd	7	18	40	30	260	300
		8.0	nd	nd	7	17	50	30	250	330
41574	2.5	0.2	0.10	nd	6	16	40	220	440	520
		4.0	nd	nd	5	19	50	190	390	500
		8.0	nd	nd	4	18	40	240	440	500
70974	4.5	0-3	0.16	1.7	5	19	30	60	310	370
		4.0	nd	nd	9	15	20	110	300	360
		7.0	nd	nd	4	15	20	70	260	280
		9.0	nd	nd	nd	25	90	30	nd	340
80874	4.3	0-3	0.16	3.2	7	19	60	20	380	430
90574	2.0	0-3	0.17	15.5	14	36	50	10	200	460
		4.0	nd	nd	12	34	100	10	250	440
		7.0	nd	nd	12	35	130	10	260	350
		9.0	nd	nd	21	42	390	50	470	750
POWERS										
40280	3.5	comp	0.11	nd	12	18	21	0	115	190
72180	3.2	0-3	0.13	3.0	3	10	13	10	250	280
QUADDICK										
41780	5.1	comp	0.10	nd	16	18	28	51	246	246
82580	2.3	0-5	0.22	4.0	7	12	100	16	386	461
QUASSAPAUG										
110573	2.2	0.2	0.15	nd	6	19	40	90	320	410
		5.0	nd	nd	6	17	40	90	320	440
		10.0	nd	nd	6	18	60	50	270	360
		15.0	nd	nd	5	20	130	50	360	430
42274	2.5	0.2	0.10	nd	3	16	10	60	260	390
		5.0	nd	nd	4	15	20	110	310	440
		10.0	nd	nd	3	16	0	80	290	410
		15.0	nd	nd	3	15	40	160	300	500
62774	6.8	0-3	0.15	3.2	6	9	30	50	250	480
		5.0	nd	nd	5	13	20	50	260	310
		10.0	nd	nd	2	14	90	110	350	570
		15.0	nd	nd	4	21	270	130	550	650
		19.0	nd	nd	nd	41	460	210	nd	970
72374	7.5	0-3	0.13	2.3	5	12	50	30	350	470
82674	6.0	0-3	0.14	3.2	14	21	30	20	290	430
		5.0	nd	nd	6	19	20	60	270	380
		10.0	nd	nd	5	12	40	70	250	340
		14.0	nd	nd	nd	26	240	150	nd	660
		18.0	nd	nd	nd	50	430	100	nd	820

APPENDIX III (continued)

Lake and date sampled	Trans., m	Depth, m	Alk., meq/l	Chl-a	Sol. P	Total P	NH_4-N	NO_3-N	Sol. N	Total N
						---ppb---				
QUONNIPAUG										
40180	3.0	comp	0.74	nd	18	27	12	111	166	251
71680	4.4	0-4	0.64	5.0	5	11	22	5	200	265
		6.0	0.60	nd	2	13	22	8	108	168
		8.0	0.70	nd	8	11	8	160	250	320
		11.0	0.61	nd	10	16	18	196	301	371
		13.0	0.72	nd	7	13	154	106	346	391
71680	nd	0-4	0.64	5.0	5	13	13	0	140	245
		6.0	0.60	nd	4	16	33	0	125	240
		8.0	0.72	nd	5	27	120	5	275	445
		10.0	1.08	nd	9	69	542	6	686	991
ROGERS										
40378	5.0	comp	nd	nd	4	10	160	100	200	340
71178	4.0	0-10	0.22	4.0	10	13	nd	110	nd	410
		14.0	nd	nd	7	18	nd	160	nd	330
		18.0	nd	nd	7	21	nd	290	nd	770
ROSELAND										
102373	nd	0.2	0.48	nd	15	24	90	390	760	900
		3.0	nd	nd	14	31	80	330	720	880
		5.0	nd	nd	12	37	120	380	720	970
50274	2.0	0.2	0.30	nd	9	30	0	500	810	950
		3.0	nd	nd	10	34	30	500	750	930
		5.0	nd	nd	10	38	30	500	800	980
71774	2.5	0-2	0.48	31.0	18	47	110	260	890	1220
		3.0	nd	nd	21	48	180	290	920	1140
		5.0	nd	nd	92	125	800	130	980	1520
82774	3.0	0-3	0.53	9.9	12	29	40	30	410	650
		5.0	nd	nd	nd	114	320	230	nd	1390
SHENIPSIT										
102573	2.5	0.2	0.15	nd	7	11	40	40	200	350
		5.0	nd	nd	5	9	50	40	160	320
		10.0	nd	nd	10	10	70	40	190	490
		15.0	nd	nd	15	923	350	60	57	680
42274	3.0	0.2	0.09	nd	12	18	40	190	380	410
		5.0	nd	nd	10	18	10	240	400	480
		10.0	nd	nd	9	17	40	190	410	420
		15.0	nd	nd	10	21	80	190	380	450
62874	4.5	0.2	0.24	5.9	5	11	20	170	350	430
		5.0	nd	nd	5	16	40	130	300	360
		10.0	nd	nd	5	11	80	260	460	500
		15.0	nd	nd	7	15	120	270	610	610
82074	3.5	0-3	0.16	5.3	3	10	0	40	400	560
		5.0	nd	nd	2	10	0	40	310	410
		10.0	nd	nd	3	10	0	240	540	540
		15.0	nd	nd	2	10	0	140	390	490
		18.0	nd	nd	nd	22	150	270	nd	870
SILVER										
32679	2.5	comp	0.73	nd	7	21	240	0	370	400
83079	1.8	0-2	1.38	23.0	69	107	22	0	nd	1100
		3.0	nd	nd	40	137	48	0	nd	1119
SQUANTZ										
41580	5.2	comp	0.34	nd	7	16	28	121	201	361
80780	3.4	0-6	0.46	6.0	12	16	16	11	181	280
		8.0	0.50	nd	13	23	24	11	171	316
		10.0	0.54	nd	5	21	203	18	383	478
		12.0	0.58	nd	5	17	320	11	436	586
		13.0	0.76	nd	15	25	550	10	710	820

APPENDIX III (continued)

Lake and date sampled	Trans., m	Depth, m	Alk., meq/l	Chl-a	Sol. P	Total P	NH$_4$-N ppb	NO$_3$-N	Sol. N	Total N
TAUNTON										
110273	4.5	0.2	0.49	nd	55	77	260	90	540	570
		4.0	nd	nd	63	84	260	90	530	610
		8.0	nd	nd	62	84	280	40	470	530
42974	3.5	0.2	0.42	nd	8	21	50	10	210	350
		4.0	nd	nd	7	24	50	10	270	420
		8.0	nd	nd	4	27	40	10	300	360
81374	3.3	0-3	0.45	5.5	6	22	40	0	410	690
		4.0	nd	nd	7	18	10	0	270	360
		8.0	nd	nd	19	151	50	0	690	870
TERRAMUGGUS										
111573	5.5	0.2	0.23	nd	8	14	30	50	270	310
		4.0	nd	nd	8	15	20	40	260	300
		8.0	nd	nd	6	12	10	40	260	260
		10.0	nd	nd	8	12	10	40	230	310
41574	4.4	0.2	0.19	nd	2	22	40	170	430	490
		4.0	nd	nd	4	22	20	180	370	420
		8.0	nd	nd	4	21	30	170	330	410
80874	6.0	0-3	0.26	2.4	3	14	60	110	470	590
		4.0	nd	nd	2	10	10	80	200	330
		8.0	nd	nd	4	25	10	60	260	390
TYLER										
41179	3.3	comp	0.60	nd	4	23	38	0	310	630
81779	3.8	0-5	1.12	7.0	6	19	20	0	438	569
		6.0	nd	nd	7	16	0	11	386	411
UNCAS										
40280	5.2	comp	0.18	nd	6	10	12	0	10	145
73080	5.4	0-4	0.20	2.0	6	10	50	4	204	254
		5-6	0.20	nd	5	14	46	0	230	285
		8.0	0.21	nd	6	18	54	6	191	286
		10.0	0.24	nd	6	15	69	15	230	300
WARAMAUG										
110573	2.0	0.2	0.41	nd	10	53	40	50	300	550
		4.0	nd	nd	7	59	20	50	230	510
		8.0	nd	nd	9	62	10	30	240	520
43074	2.0	0.2	0.34	nd	2	25	50	80	220	470
		4.0	nd	nd	3	28	40	80	220	460
		8.0	nd	nd	2	28	90	80	270	490
70274	2.3	0.2	0.38	13.1	7	24	0	70	370	600
		4.0	nd	nd	9	36	0	20	370	510
		8.0	nd	nd	nd	115	310	150	nd	840
82274	3.2	0-3	0.43	9.0	12	24	40	70	410	670
		4.0	nd	nd	10	21	30	30	260	430
		8.0	nd	nd	nd	57	240	180	nd	980
WAUMGUMBAUG										
40880	3.6	comp	0.31	nd	5	16	33	84	314	424
80580	6.1	0-3	0.24	2.0	5	10	16	6	316	416
		7.0	0.34	nd	7	20	20	10	370	485
		9.0	0.42	nd	10	29	35	10	340	555

APPENDIX III (continued)

Lake and date sampled	Trans., m	Depth, m	Alk., meq/l	Chl-a	Sol. P	Total P	NH$_4$-N	NO$_3$-N	Sol. N	Total N
							ppb			
WEST HILL										
92673	nd	0.2	nd	nd	7	9	40	30	230	280
		10.0	nd	nd	4	10	40	20	240	300
		14.0	nd	nd	4	10	200	70	420	420
42374	5.5	0.2	0.13	nd	3	9	30	70	190	260
		5.0	nd	nd	3	8	30	60	270	270
		10.0	nd	nd	4	10	20	20	220	210
		14.0	nd	nd	4	7	0	20	160	310
70274	6.8	0.2	0.16	1.8	5	8	20	30	180	190
		5.0	nd	nd	3	8	20	30	160	200
		10.0	nd	nd	3	9	0	50	190	240
		15.0	nd	nd	3	14	20	50	240	310
82274	7.2	0-3	0.13	1.2	6	6	30	70	250	300
		5.0	nd	nd	3	9	40	30	210	280
		10.0	nd	nd	4	14	60	30	170	280
		14.0	nd	nd	nd	16	50	50	nd	370
WEST SIDE										
42878	nd	comp	0.75	nd	4	14	0	100	420	660
72078	4.0	0-5	1.22	3.0	10	11	nd	0	380	530
		8.0	nd	nd	13	38	nd	0	450	450
WINNEMAUG										
41180	1.3	comp	0.38	nd	10	32	71	328	663	1003
81580	1.3	0-2	0.56	24.0	8	32	87	0	535	1020
		3.5	0.58	nd	8	32	113	0	590	1025
		4.0	0.86	nd	7	48	1060	3	1433	1893
WONONPAKOOK										
42180	1.4	comp	2.86	nd	18	43	50	51	386	576
71880	2.1	0-3	2.36	15.0	9	28	65	0	405	670
		4.0	2.35	nd	8	32	55	0	410	670
		5.0	2.65	nd	8	53	55	0	310	680
		6.0	2.88	nd	10	72	316	0	460	935
WONONSCOPOMUC										
101773	4.3	0.2	2.00	nd	7	22	10	30	320	360
		5.0	nd	nd	8	18	10	10	330	400
		10.0	nd	nd	8	107	30	10	290	780
		15.0	nd	nd	12	30	50	10	330	390
		25.0	nd	nd	350	355	1450	60	1610	1810
50774	1.0	0.2	2.50	nd	5	50	30	50	310	740
		5.0	nd	nd	3	52	10	40	330	810
		10.0	nd	nd	4	44	30	40	240	740
		15.0	nd	nd	7	40	20	40	300	600
		25.0	nd	nd	7	29	30	40	390	520
71174	7.3	0-3	2.18	0.7	2	12	20	40	360	400
		5.0	nd	nd	1	13	20	40	380	460
		10.0	nd	nd	5	123	0	30	190	1160
		15.0	nd	nd	4	31	10	40	400	560
		20.0	nd	nd	10	18	130	40	400	490
		25.0	nd	nd	nd	192	530	50	nd	990
90474	8.2	0-3	2.07	2.4	7	16	40	10	270	620
		5.0	nd	nd	26	54	240	10	490	700
		10.0	nd	14.6	5	159	0	10	350	1400
		15.0	nd	nd	5	25	0	10	240	300
		20.0	nd	nd	62	74	390	10	640	610
		25.0	nd	nd	276	296	870	0	1260	1380
WYASSUP										
40479	4.3	comp	0.10	nd	6	16	16	0	220	250
81679	4.3	0-5	0.18	3.0	6	10	38	19	369	482
		6.0	nd	nd	5	7	38	0	238	275
		8.0	nd	nd	2	11	28	0	113	300

APPENDIX III (continued)

Lake and date sampled	Trans., m	Depth, m	Alk., meq/l	Chl-a	Sol. P	Total P	NH$_4$-N	NO$_3$-N	Sol. N	Total N
							----ppb----			
ZOAR										
100473	1.5	0.2	2.02	nd	33	102	20	420	nd	1420
		5.0	nd	nd	34	52	110	430	nd	950
		9.0	nd	nd	47	62	120	nd	nd	nd
		15.0	nd	nd	44	62	100	530	nd	950
53174	2.2	0.2	1.53	nd	32	65	160	310	700	810
		5.0	nd	nd	35	63	160	240	590	690
		10.0	nd	nd	36	80	170	230	600	700
		15.0	nd	nd	40	68	160	210	590	650
70374	1.0	0.2	1.67	103.0	12	77	40	50	240	1090
		3.0	nd	nd	23	62	170	190	460	740
		5.0	nd	nd	18	49	160	190	510	680
		10.0	nd	nd	19	37	160	290	670	680
		15.0	nd	nd	32	58	210	390	740	810
		20.0	nd	nd	nd	84	800	100	nd	1140
73174	2.5	0-3	1.81	25.0	24	60	0	320	680	960
82974	2.1	0-3	1.80	35.0	10	62	40	60	400	670
		5.0	nd	nd	34	85	180	120	420	680
		10.0	nd	nd	75	138	370	230	790	970
		15.0	nd	nd	nd	207	590	110	nd	900
		20.0	nd	nd	nd	351	1150	110	nd	1650

APPENDIX IV

Bathymetric data and descriptive narratives were taken principally from the report of the Connecticut State Board of Fisheries and Game (1959) and from the inventory compiled by the Connecticut Department of Environmental Protection (1980). Minor differences in lake size and depth exist between these data and those in Appendix II due to different units and methods of calculation. The data shown for temperature and dissolved oxygen are representative of mid-summer conditions but vary due to the weather during any particular year. Data on trout stocking were taken from a report of the Connecticut Department of Environmental Protection (1980-1981). Information concerning the extent and composition of major weed beds in lakes not shown in Table 1 was taken from surveys on file with the Connecticut Department of Environmental Protection. A key to common and scientific names of aquatic weeds is given below.

KEY TO AQUATIC WEEDS

arrowhead (duck potato)	*Sagittaria latifolia*
big lea pondweed	*Potamogeton amplifolius*
bladderwort	*Utricularia sp.*
bur reed	*Sparganium sp.*
bushy pondweed (water nymph)	*Najas flexilis*
cattail	*Typha sp.*
coontail (hornwort)	*Ceratophyllum demersum*
curly pondweed	*Potamogeton crispus*
elodea (waterweed)	*Elodea canadensis*
fanwort	*Cabomba caroliniana*
fern pondweed	*Potamogeton sp.*
goldenpert	*Gratiola aurea*
leafy pondweed	*Potamogeton foliosus*
lesser duckweed	*Lemna minor*
little floating heart	*Nymphoides cordata*
muskgrass	*Chara sp.*
pickerelweed	*Pontederia cordata*
pipewort	*Ericaulon septangulare*
red head grass	*Potamogeton perfoliatus*
Robbins pondweed	*Potamogeton robbinsii*
snailseed pondweed	*Potamogeton diversifolius*
spatterdock	*Nuphar sp.*
spikerush	*Eleocharis sp.*
stonewort	*Nitella flexilis*
water bulrush	*Scirpus sp.*
watermeal	*Wolffia columbiana*
water milfoil	*Myriophyllum sp.*
watershield	*Brasenia schreberi*
white water buttercup	*Ranunculus aquatilis*
white water lily	*Nymphaea odorata*
wild celery	*Vallisneria americana*

ALEXANDER LAKE

Alexander Lake is located in Windham County in the town of Killingly. It has a surface area of 190.4 acres, a maximum depth of 53 feet and an average depth of 24.2 feet.

The 761.6-acre watershed is 9.8% urban, 7.5% agricultural and 82.7% wooded or wet. Topographic variation in the watershed is relatively slight. The greatest change occurs along the southwestern portion where the landscape rises 110 feet above the lake. The outlet is an unnamed stream on the southern shore that joins the Quinebaug River 1½ miles to the south.

Alexander Lake has been stocked with adult brown and rainbow trout in recent years.

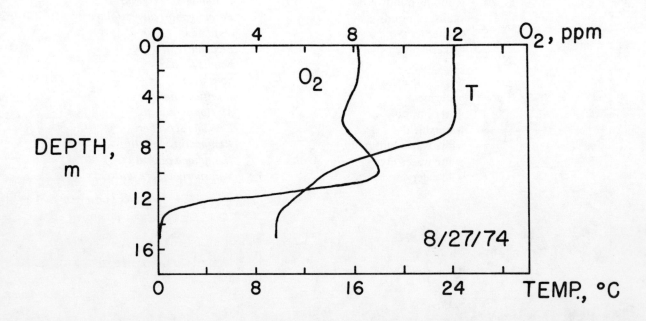

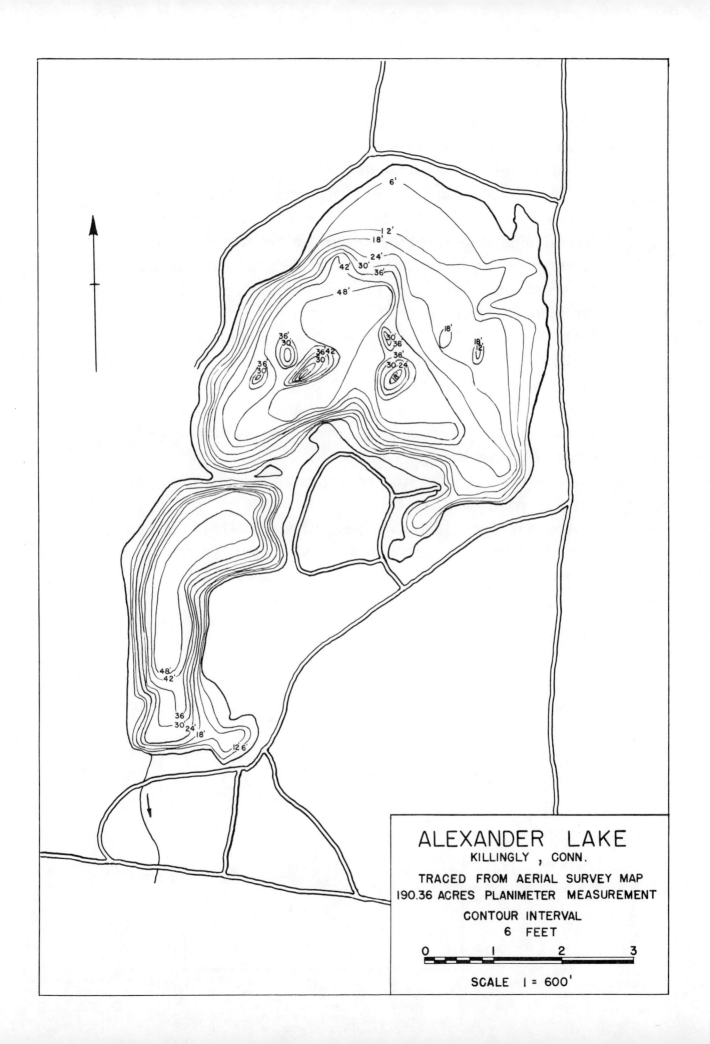

AMOS LAKE

Amos Lake is located in New London County in the town of Preston. It has a surface area of 105.1 acres, a maximum depth of 48 feet and an average depth of 19.1 feet. The level has been raised several feet by a small dam.

The 537.6-acre watershed is 4.2% urban, 11.3% agricultural, and 84.5% wooded or wet. Severe slopes caused by a ridge, several hummocks along the eastern and western shore, and several wetlands contiguous with the lake's surface characterize the shore. Generally, severe to moderate slopes prevail throughout the watershed.

On August 18, 1980, aquatic weeds were abundant in the shallow shore areas. Mixed emergent weeds included white water lily, spatterdock, and pickerelweed. Submerged weeds included pipewort, wild celery, fern pondweed, and numerous large patches of filamentous algae. A cove on the southeastern shore contained dense beds of bladderwort.

Amos Lake has been stocked with adult brown and rainbow trout in recent years.

Access is from Rte. 164 on the southwestern shore through a State boat launching area and for residents at town-operated facilities.

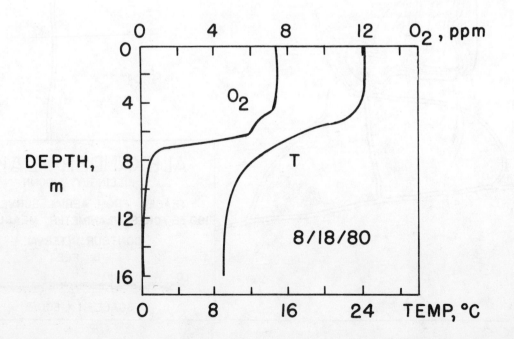

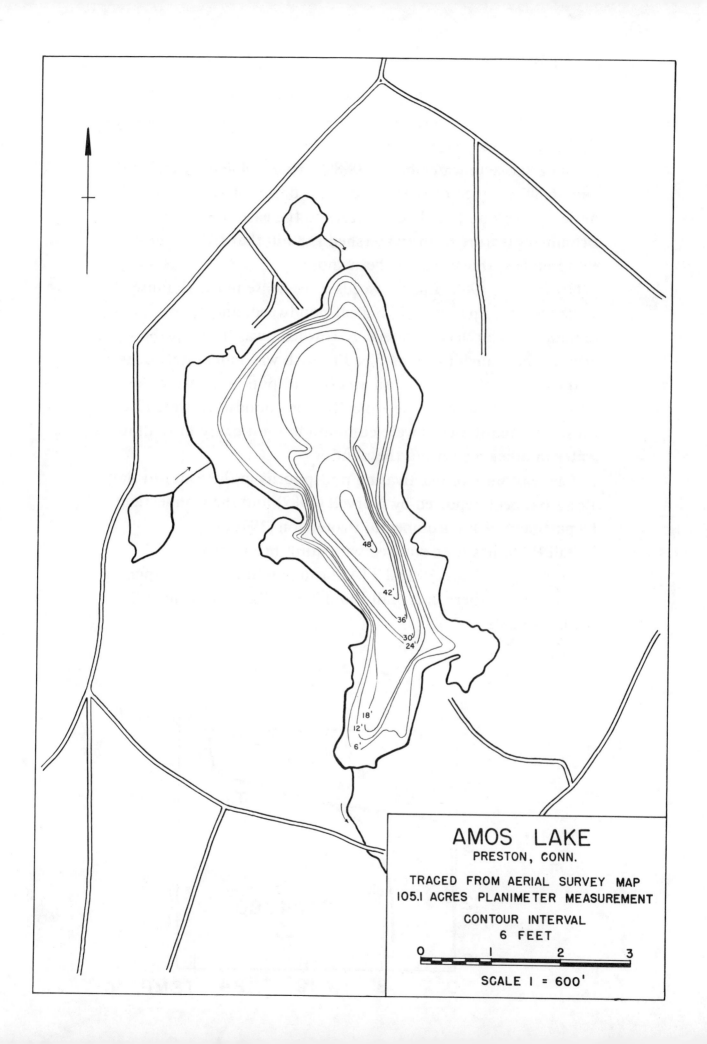

BALL POND

Ball Pond is located in Fairfield County in the town of New Fairfield. It has a surface area of 89.9 acres, a maximum depth of 52 feet and an average depth of 22.7 feet.

The 236.8-acre watershed is 38.8% urban, 14.4% agricultural, and 46.8% wooded or wet. No streams drain into the pond. The outlet, Ball Pond Brook, discharges to the east. The most prominent feature of the watershed is a hill that rises several hundred feet above the northern shore.

On July 24, 1980, aquatic weeds were dense in the shallow southern end. Spatterdock, bigleaf pondweed, and coontail were abundant to depths of 6.5 feet. Arrowhead was dense at the shore around the launching area. Weed growth was moderately dense in the northern end. Coontail and bigleaf pondweed were the most common, while elodea and Robbins pondweed were also present. Aquatic weeds were common but not dense in shallow water in other sections of the pond.

The feasibility of inactivating phosphorus in the water in Ball Pond has been reported by Norvell (1982) and the Connecticut Department of Environmental Protection (1979a).

Ball Pond has been stocked with adult brown trout.

Access is off Rtes. 39 and 37A through a State boat launching area at the southern end. Ball Pond Town Park offers access to town residents.

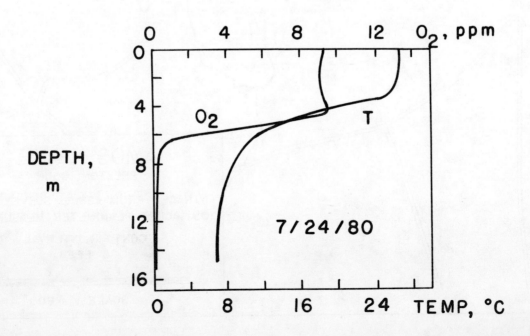

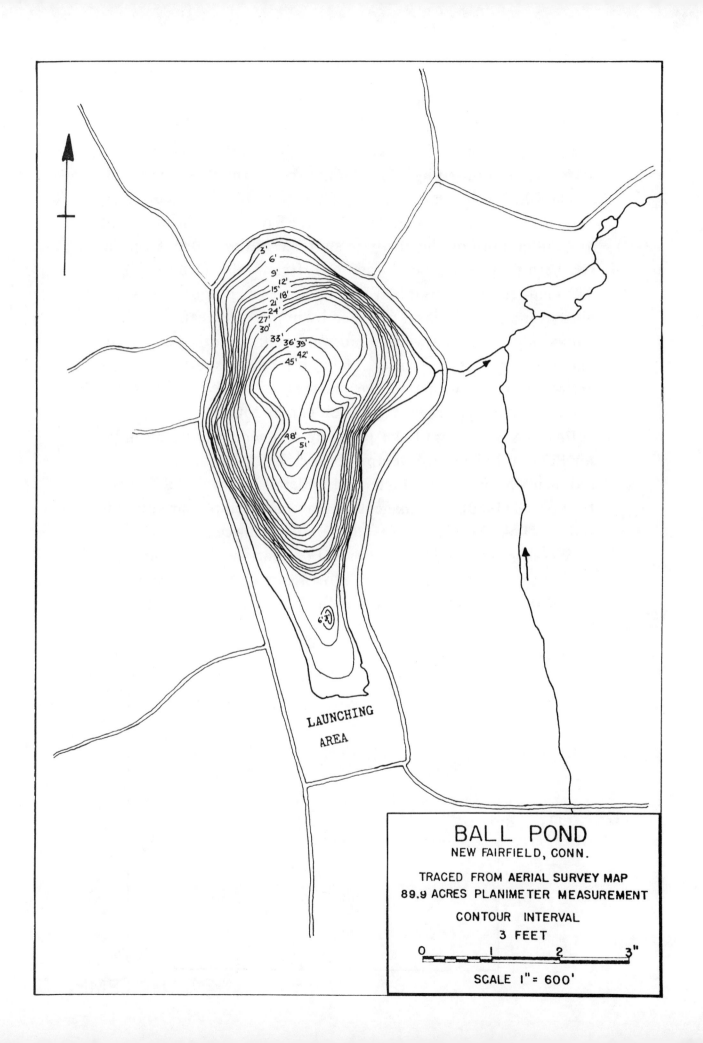

BANTAM LAKE

Bantam Lake is located in Litchfield County in the towns of Litchfield and Morris. It has a surface area of 916 acres, a maximum depth of 25 feet and an average depth of 14.3 feet. The level has been raised by a low dam at the northwestern end.

The 20,218-acre watershed is 6.8% urban, 13.7% agricultural, and 79.5% wooded or wet. The Bantam River is the primary inlet and outlet, both on the northern shore. Whittlesey Brook on the southern shore and several unnamed streams also feed the lake. The character of the Bantam River changes as it approaches the ancient limits of the lake, reflected by numerous meanders commonly associated with an old watercourse. Both the eastern and western shores are rocky. The northern and southern shores are relatively flat with extensive shoal areas where considerable submerged and emergent vegetation can be found.

Portions of Bantam Lake are being dredged. A nutrient budget for Bantam Lake was developed by Frink (1967), and its sediments have been characterized in detail by Frink (1969) and by Norvell (1980). A management plan has been prepared by the Bantam Lake Watershed Task Group Committee (1981).

Access is at the White Memorial Foundation boat launching area at the northern end, at the Morris town boat launching area, and at the Sandy Beach boat launching area at the southern end.

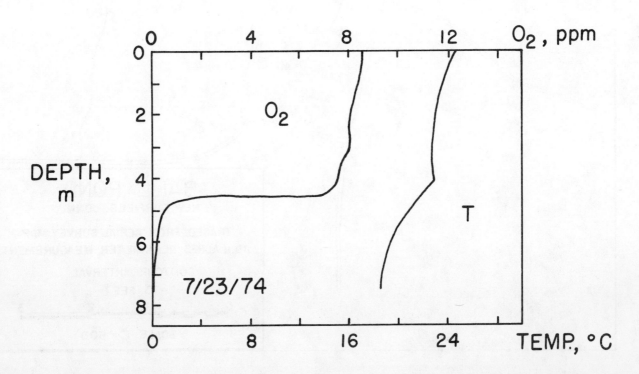

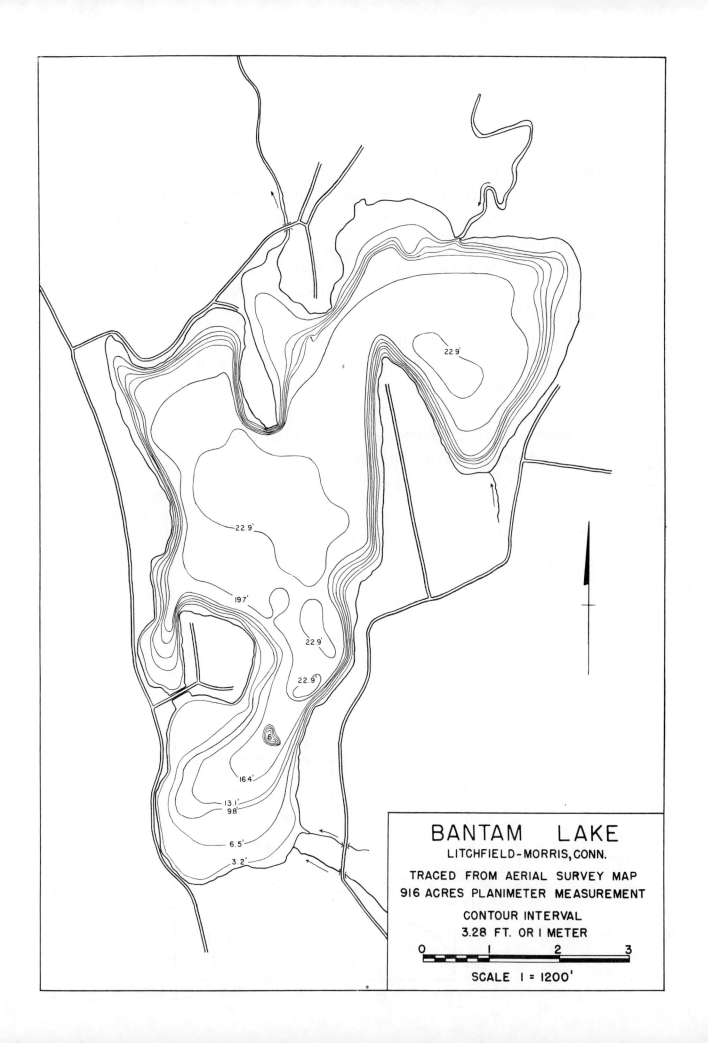

BASHAN LAKE

Bashan Lake is located in Middlesex County in the town of East Haddam. It has a surface area of 276.3 acres, a maximum depth of 48 feet and an average depth of 15.9 feet. The level has been raised by a 25-foot masonry and earthen dam. The lake is owned by the Moodus Water Company. Water is drawn from it to maintain the Moodus Reservoir, causing moderate fluctuation of the level of the lake.

The 1,274-acre watershed is 23.0% urban, 9.0% agricultural and 68.0% wooded or wet. The southeastern shore has severe slopes.

On July 22, 1980, aquatic weeds were moderately dense in a few areas. Shallow coves contained several species of pondweeds including Robbins pondweed, bladderwort, and a mixture of reed grasses which included bur reed and water bulrush.

Access is through a State boat launching area off Rte. 82 at the southeastern end of the lake.

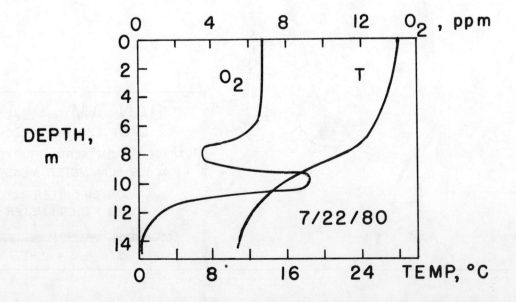

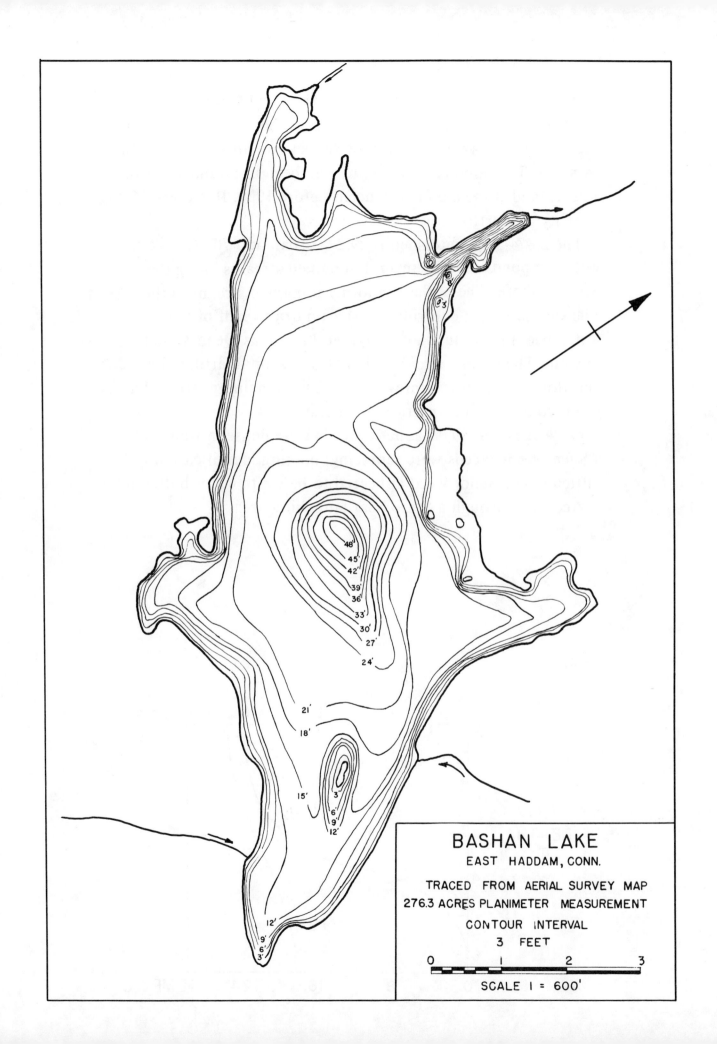

BATTERSON PARK POND

Batterson Park Pond is located in Hartford County in the towns of Farmington and New Britain. It has a surface area of 162.7 acres, a maximum depth of 20 feet and an average depth of 14.8 feet. The pond is artificial, having been impounded by an earthen and stone dam constructed before 1900. It is owned by the City of Hartford.

The 2,874-acre watershed is 29.2% urban and 70.8% wooded or wet. The pond is fed by several unnamed streams along the western shore. The outlet is another unnamed stream to the east. The southernmost contributing stream drains half of the watershed, including a large wetland known as Dead Wood Swamp. The middle stream originates at a small artificial pond and flows through several wetlands. The remaining stream was rerouted during construction of Interstate 84.

On August 29, 1979, aquatic weeds were dense in water 1.6 to 4.8 feet deep. Weeds were senescing, making identification difficult. The major weed appeared to be white water buttercup.

Access is through a State boat launching area.

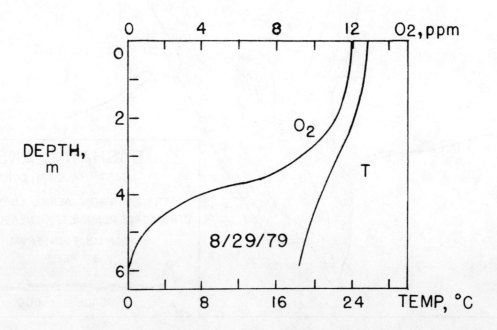

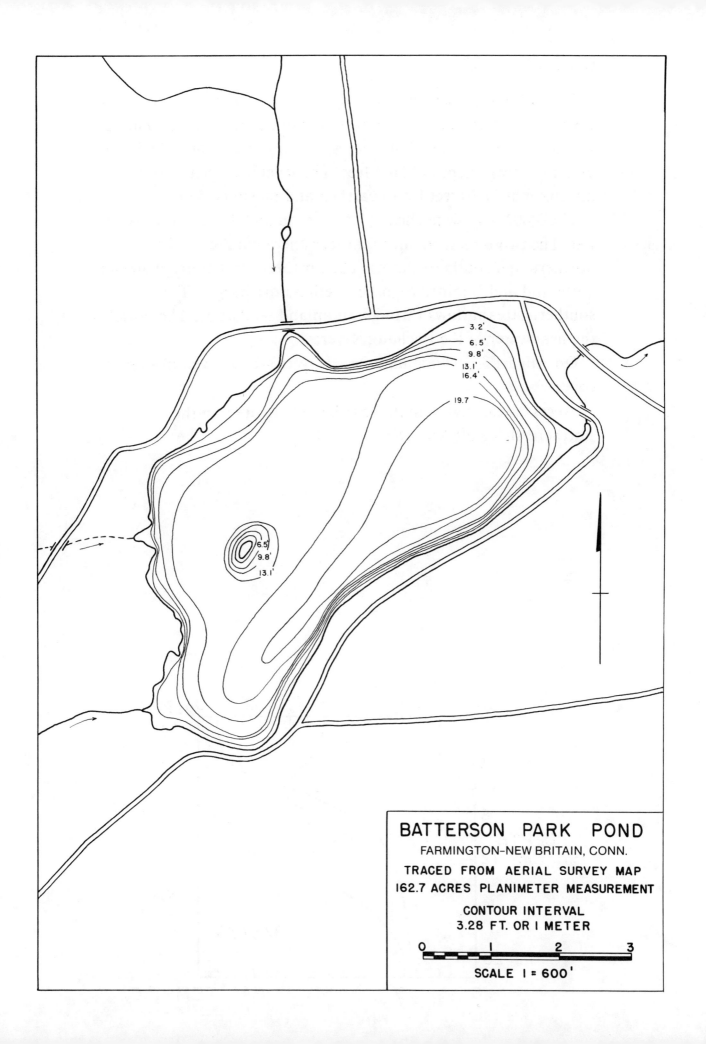

BEACH POND

Beach Pond is located primarily in New London County in the town of Voluntown. A small portion extends into Rhode Island. It has a surface area of 394.3 acres, a maximum depth of 65 feet and an average depth of 20.1 feet. The level has been raised approximately 10 feet by an earthen and masonry dam.

The 3,002-acre watershed is 2.5% urban and 97.5% wooded or wet. The two largest streams that feed the pond are on the northern and southern shores. The northern stream originates in a wetland and is joined by other wetland discharges. The southern stream flows through an upland landscape. The outlet is the headwater of the Pachaug River.

On August 3, 1979, aquatic weeds were scarce, even in shallow coves.

Access is through a State boat launching area on the northern shore off Rte. 165.

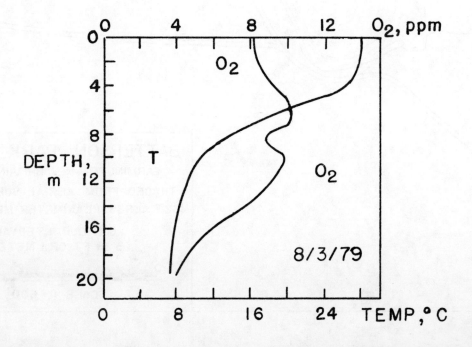

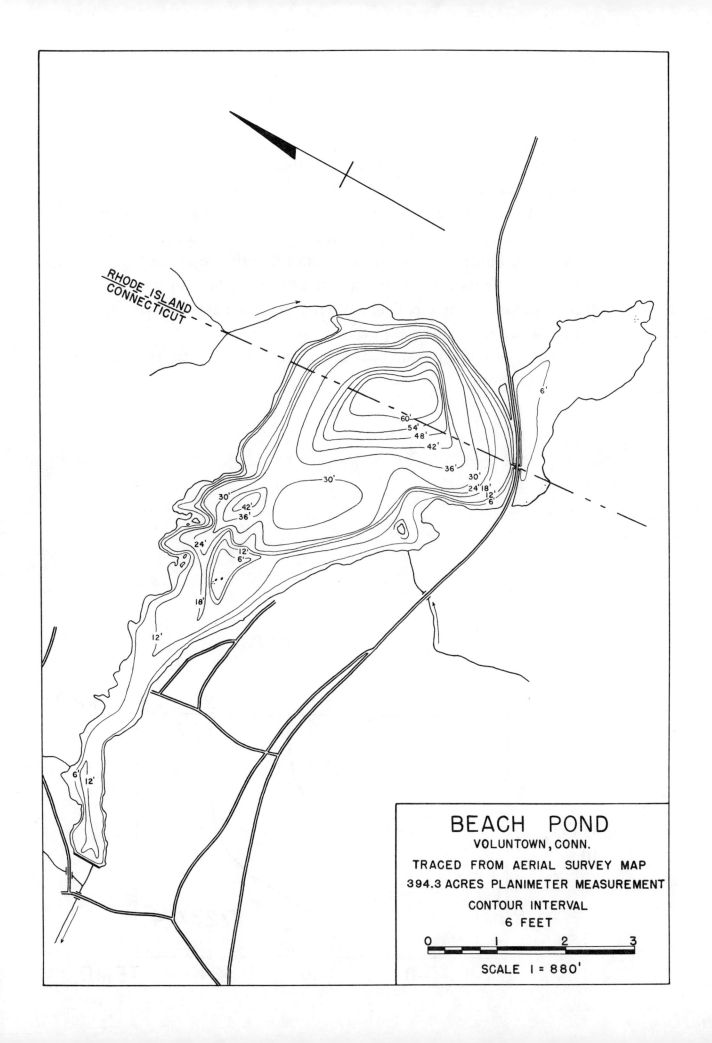

BESECK LAKE

Beseck Lake is located in Middlesex County in the town of Middlefield. It has a surface area of 119.6 acres, a maximum depth of 24 feet and an average depth of 11.2 feet.

The 1,344-acre watershed is 12.8% urban, 14.5% agricultural and 72.7% wooded or wet.

The lake is artificial. It is drained by an unnamed stream to the east which merges with the Coginchaug River. The eastern slope of Beseck Mountain comprises a substantial portion of the watershed. The masonry-earthen dam allows the lake to be drained almost completely. The water level fluctuates slightly due to drawdown for industrial purposes.

Access is through a State boat launching area off Rte. 147 on the eastern shore.

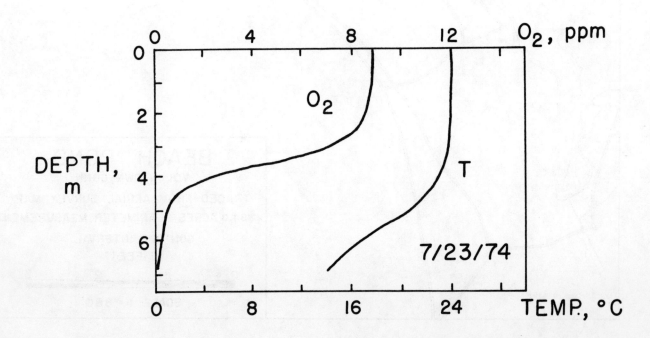

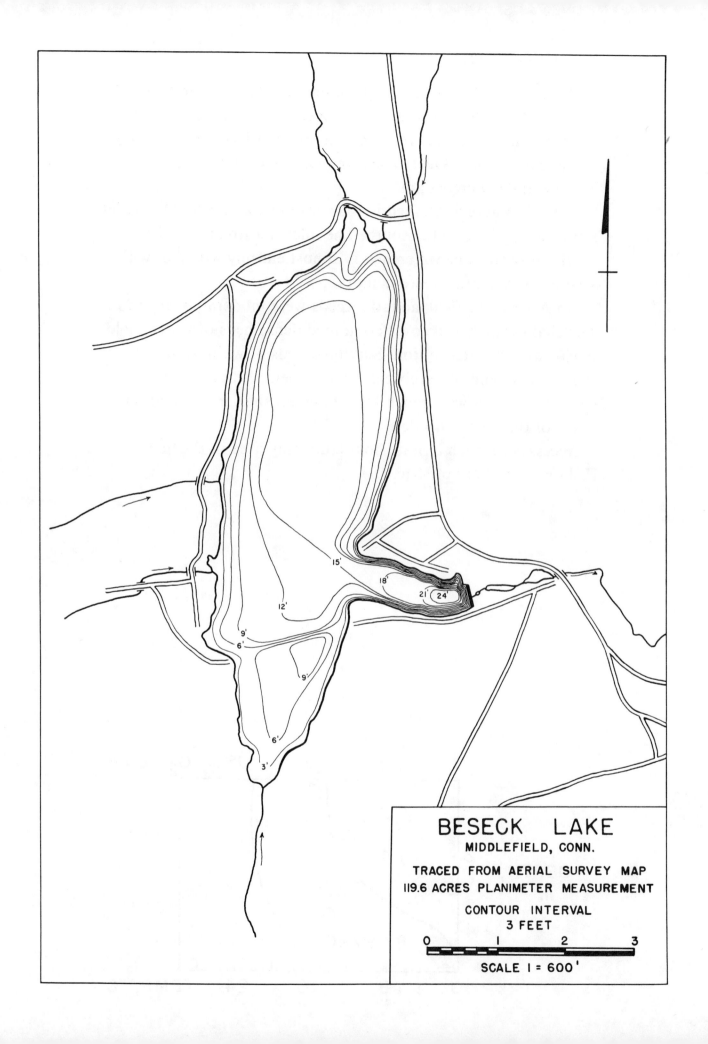

BIGELOW POND

Bigelow Pond is located in Tolland County in the town of Union. The State-owned pond in Bigelow Hollow State Park has a surface area of 18.5 acres, a maximum depth of 16 feet and an average depth of 7.5 feet. The level has been raised by an 8-foot earthen and concrete dam.

The 870.4-acre watershed is 100% wooded and wet. The outlet is a small wetland to the south. The inlet is a stream on the northern shore. The watershed is almost entirely wooded, with severe slopes on the western shore.

On August 19, 1980, aquatic weeds were moderately abundant. The shallow northern coves contained dense beds of watershield, bur reed, and water milfoil. Scattered beds of white water lily, spatterdock, and pickerelweed were present. Weeds in the southern section were much less abundant than in the shallow coves of the northern side.

Access is through a State boat launching area in Bigelow Hollow State Park off Rte. 197.

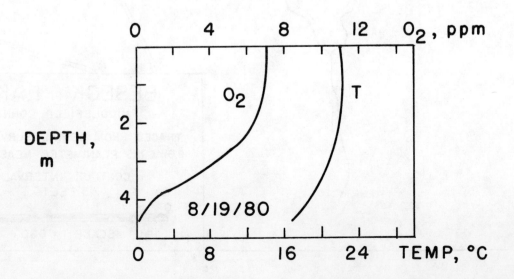

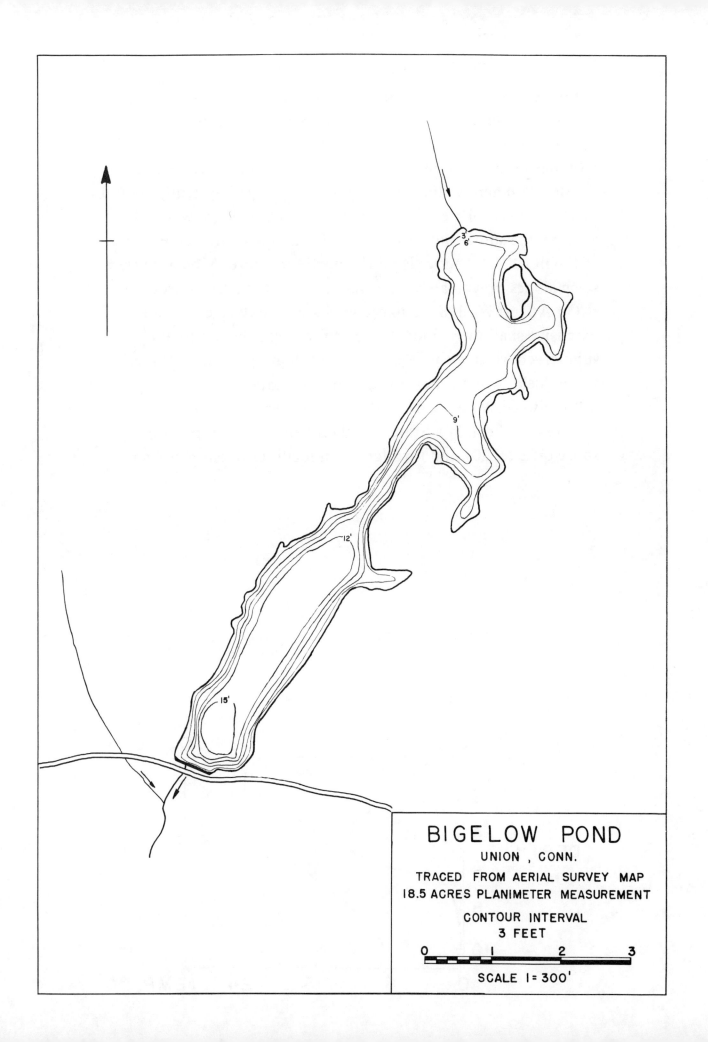

BILLINGS LAKE

Billings Lake is located in New London County in the town of North Stonington. It has a surface area of 105.1 acres, a maximum depth of 33 feet and an average depth of 13.7 feet. The level has been raised by a 10-foot masonry and concrete dam.

The 448.0-acre watershed is 5.0% urban, 3.8% agricultural and 91.2% wooded or wet. There are two relatively large wetland areas to the south and east of the lake, and a small wetland area to the north. The outlet is on the northern shore. Moderate to steep slopes prevail along the western and northern shores.

On August 20, 1980, aquatic weeds were extremely dense in extensive shallow regions. Water milfoil was the dominant submerged weed. Other emergent and floating weeds included watershield, white water lily, and occasional stands of pickerelweed. Weed beds extended to water depths of 9.8 feet.

Access is through a State boat launching area on the northern shore off Rte. 201 and through town facilities open to residents.

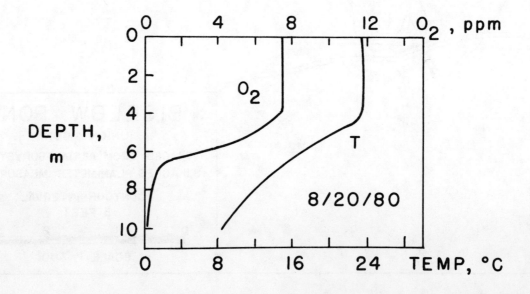

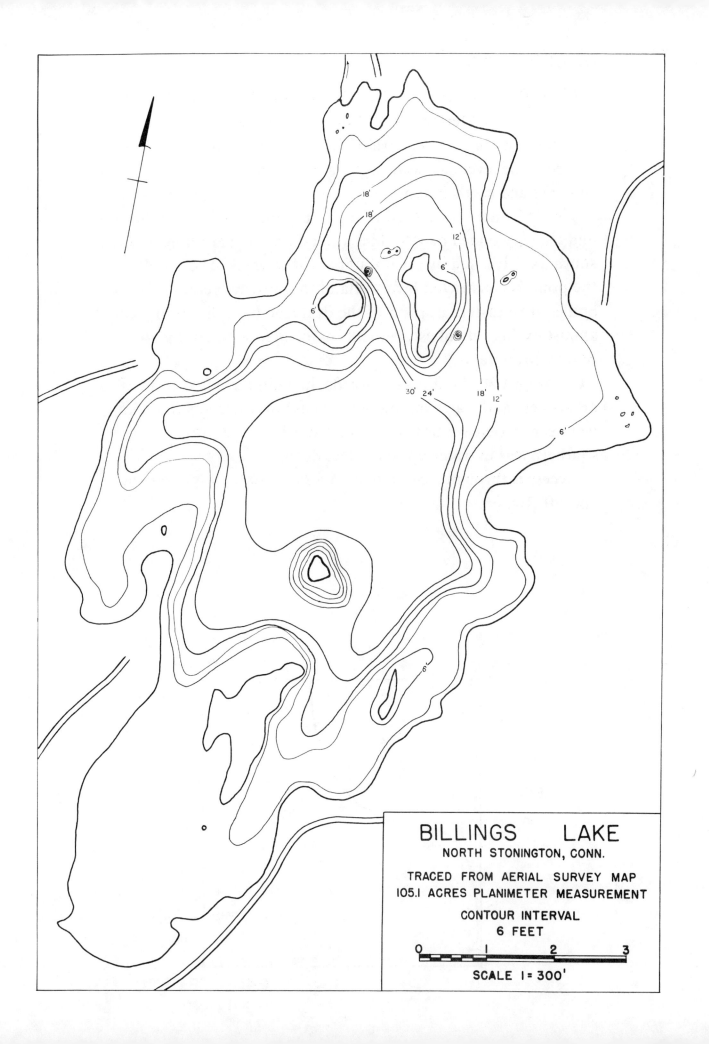

BLACK POND

Black Pond is located in New Haven and Middlesex Counties in the towns of Meriden and Middlefield. It has a surface area of 75.6 acres, a maximum depth of 23 feet and an average depth of 8.6 feet. The level has been raised approximately 5 feet by a concrete and earthen dam.

The 755.2-acre watershed is 16.7% urban, 14.0% agricultural and 69.3% wooded or wet. The watershed is linear along a north-south axis. It is fed by several unnamed streams. The middle of the pond is constricted by a wetland that nearly creates two water bodies. The eastern portion of the drainage basin is comprised almost exclusively of escarpments associated with Higby and Beseck Mountains.

On August 30, 1979, mixed weeds, including elodea, coontail, pondweed, and water milfoil, were dense in water less than 6.5 feet deep. Extensive beds of white water lily were present at the southern end in water up to 9.8 feet deep.

Access is through a State boat launching area on the northern end off Rte. 66.

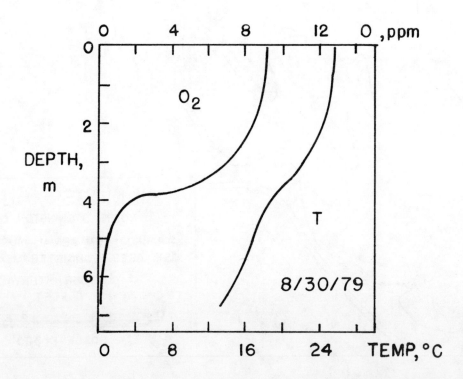

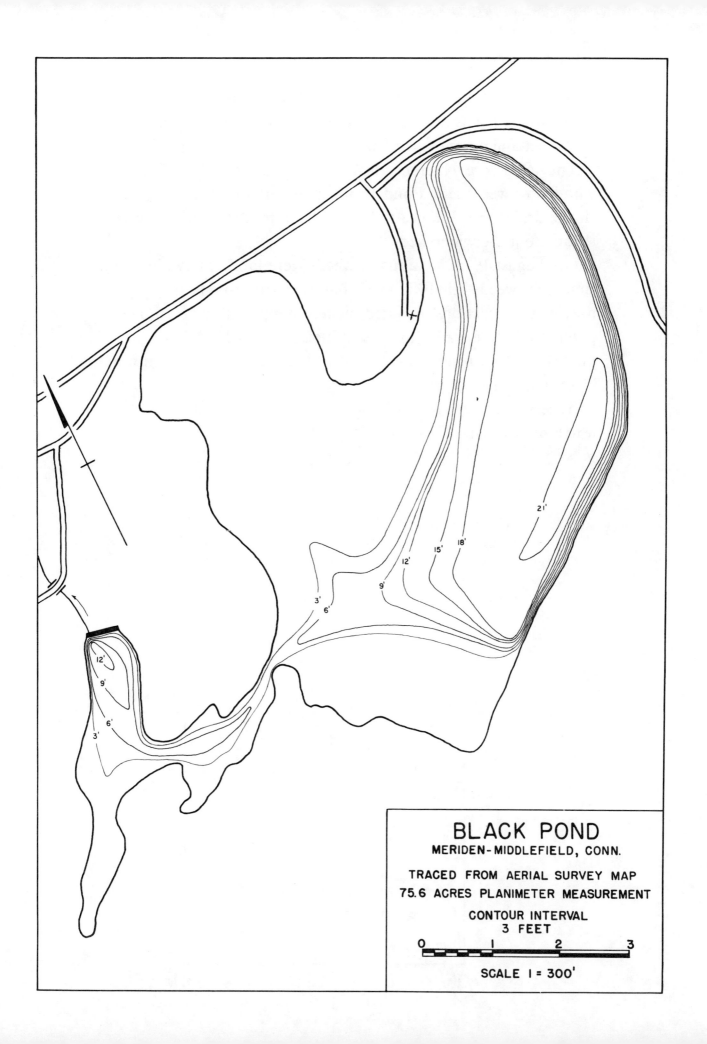

BLACK POND

Black Pond is located in Windham County in the town of Woodstock. It has a surface area of 73.4 acres, a maximum depth of 23 feet and an average depth of 12.2 feet.

The 460.8-acre watershed is 4.3% agricultural and 95.7% wooded or wet. Black Pond Brook, the outlet on the southern shore, drains into Bungee Lake. Slight to moderate slopes prevail throughout the watershed.

On August 19, 1980, aquatic weeds were moderately abundant. Emergent weeds inhabiting the shallow northern and northeastern sections included pickerelweed and bur reed. Dense patches of white water lily, watershield, spatterdock, and pipewort were present along most of the rocky shoreline in water less than 3.2 feet deep.

Access is through a State boat launching area about two miles north of Woodstock Valley off Rte. 198.

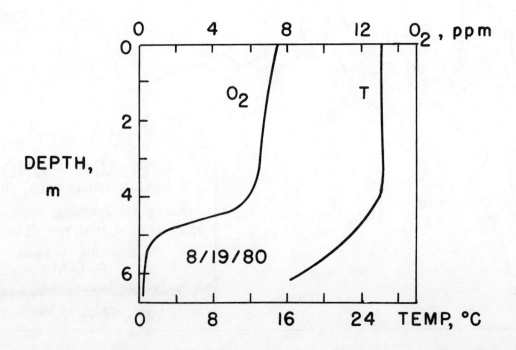

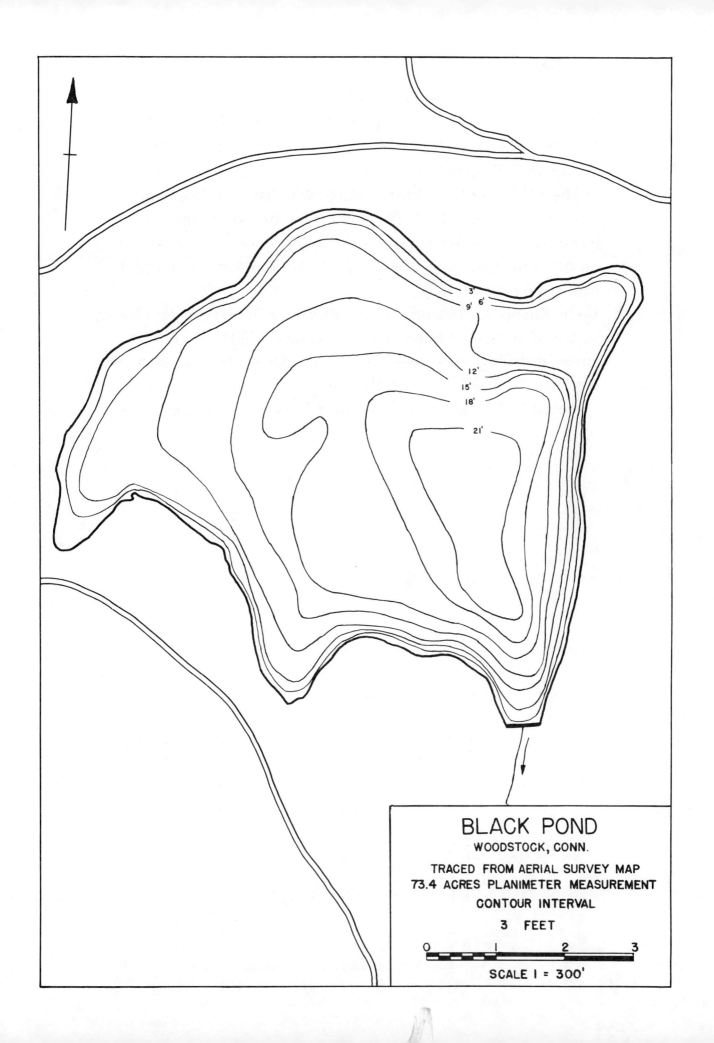

BURR POND

Burr Pond is located in Litchfield County in the town of Torrington. This State-owned artificial pond in Burr Pond State Park has a surface area of 85 acres, maximum depth of 13 feet and an average depth of 5.1 feet.

The 844.8-acre watershed is 48.6% wooded. The remainder has not yet been inventoried. The pond is fed by two permanent tributary streams and several intermittent streams. The outlet is on the eastern shore. Moderate to severe slopes are found to the south and east.

On August 8, 1980, aquatic weeds were scarce. Emergent weeds including watershield and cattail were found in the northwest coves in shallow water less than 3.2 feet deep. The boat launching area was bordered by pickerelweed.

Access is through Burr Pond State Park on the western side.

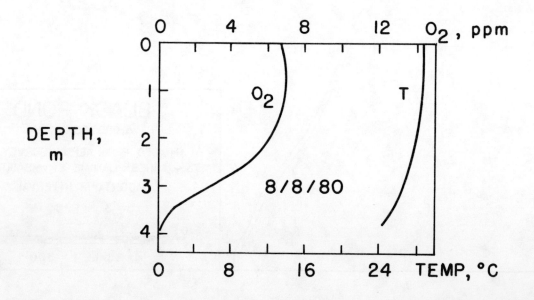

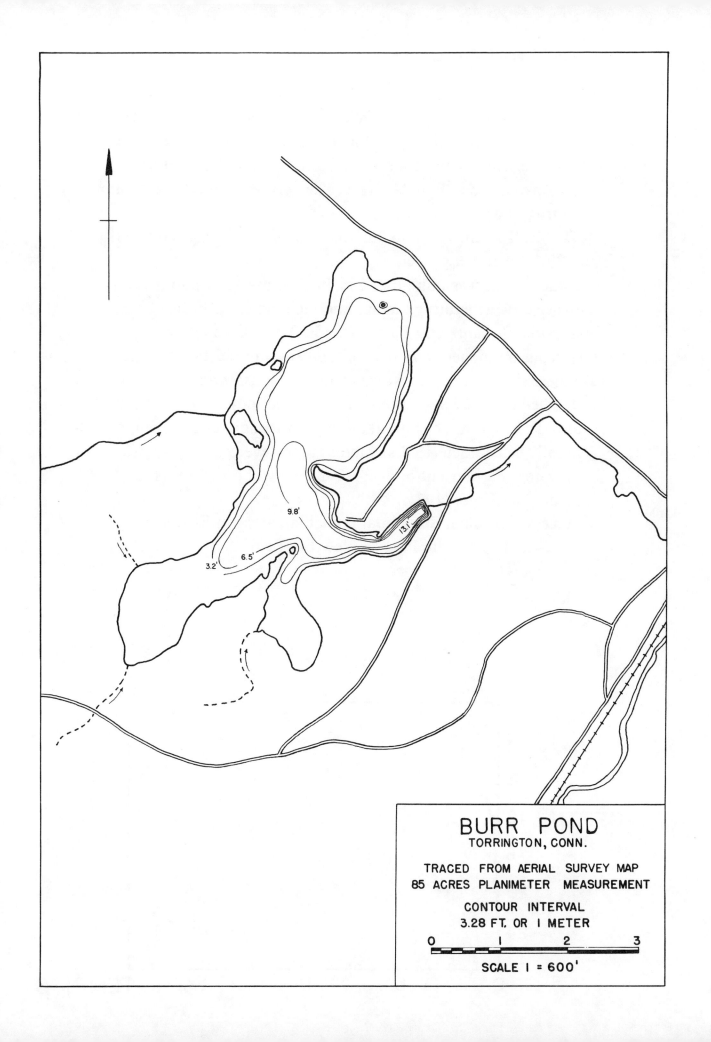

CANDLEWOOD LAKE

Candlewood Lake, the largest impoundment in the State, is located in Fairfield and Litchfield Counties in the towns of Sherman, New Milford, New Fairfield, Danbury, and Brookfield. It has a surface area of 5,420 acres, a maximum depth of 85 feet and an average depth of 29.3 feet. The earthen and concrete dam is approximately 90 feet high.

The 25,920-acre watershed is 18.7% urban, 1.2% agricultural and 80.1% wooded or wet.

Candlewood was impounded in 1923 as part of a pump storage hydroelectric generating facility and includes in its basin Neversink, Squantz and Barses Ponds. The lake is fed by numerous permanent and intermittent streams, surface runoff and periodic pumping of water from the Housatonic River. The watershed is varied and accented by several mountains that rise dramatically from the shoreline. Slopes throughout the basin can generally be characterized as moderate to severe.

A water and nutrient budget for Candlewood Lake was estimated by Frink (1971).

Access is through State boat launching areas in Danbury and at Squantz Pond State Park in New Fairfield.

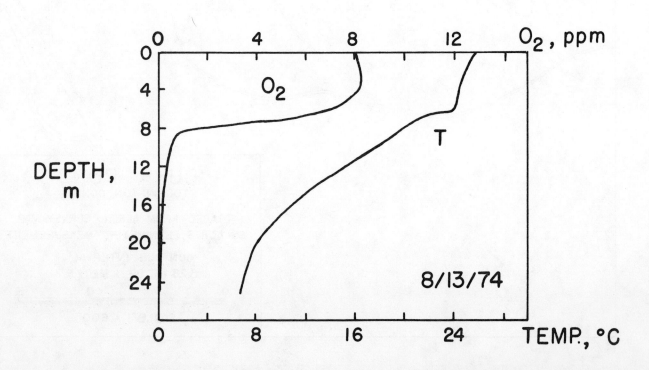

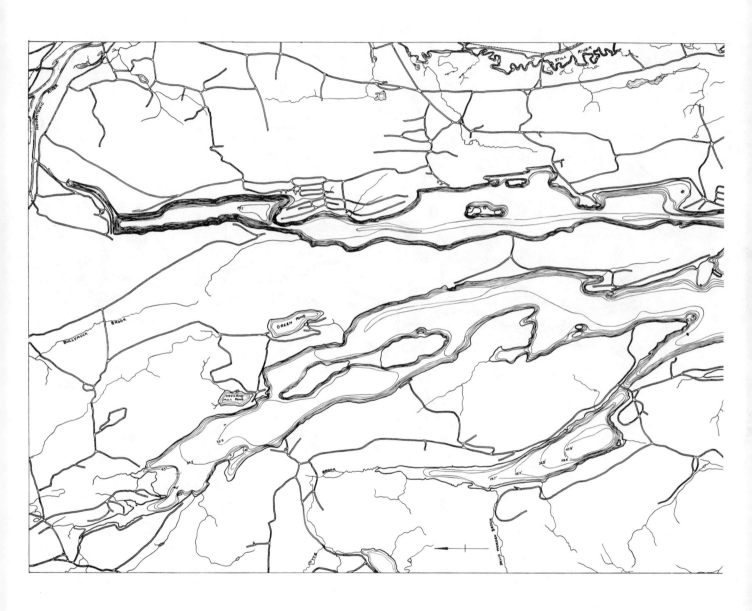

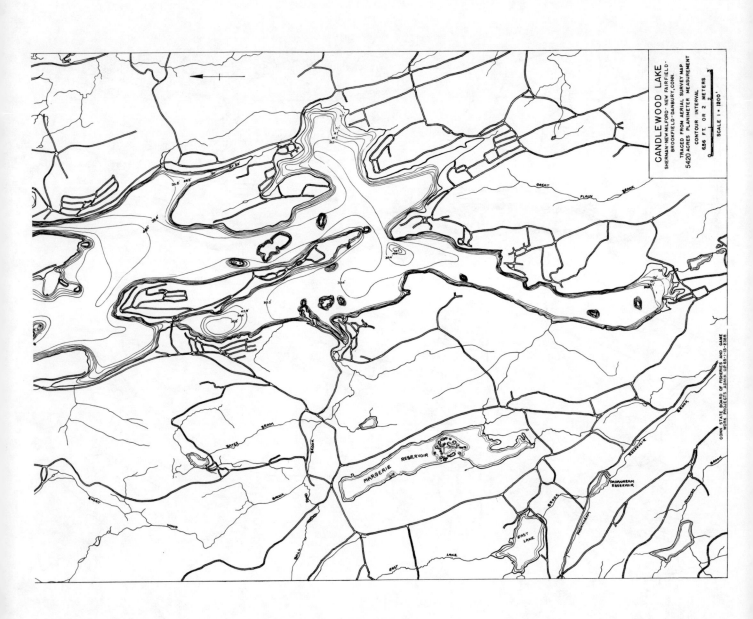

CEDAR LAKE

Cedar Lake is located in Middlesex County in the town of Chester. It has a surface area of 68 acres, maximum depth of 45 feet and an average depth of 19.3 feet. The level has been raised several feet by a low dam.

The 2,067-acre watershed is 3.6% urban, 0.5% agricultural and 95.9% wooded or wet. The inlets are Pattaconk Brook and a stream on the northern shore. The outlet is a wetland, Cedar Swamp, to the south. Two streams drain Cedar Swamp, Pattaconk Brook to the east and Burr Brook to the west. The landscape of the upper basin is moderately undulating, becoming steeper closer to the lake with a mostly wooded eastern shoreline and a mostly open western shoreline.

On August 15, 1979, aquatic weeds were dense only near the inlets on the southern and northwestern shores. Water milfoil and white water lily were the most common. White water lily, spatterdock, watershield and pickerelweed were sparsely present in many areas less than 6.5 feet deep.

Cedar Lake has been stocked with adult brown and rainbow trout in recent years.

Access is through a State boat launching area off Rte. 148 at the northern end.

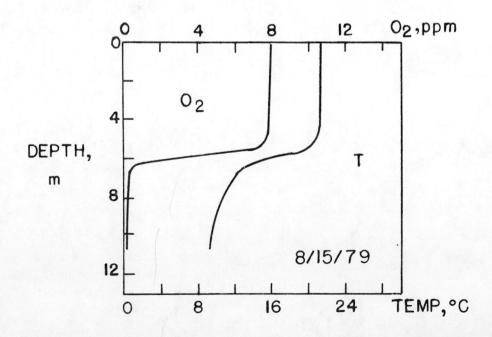

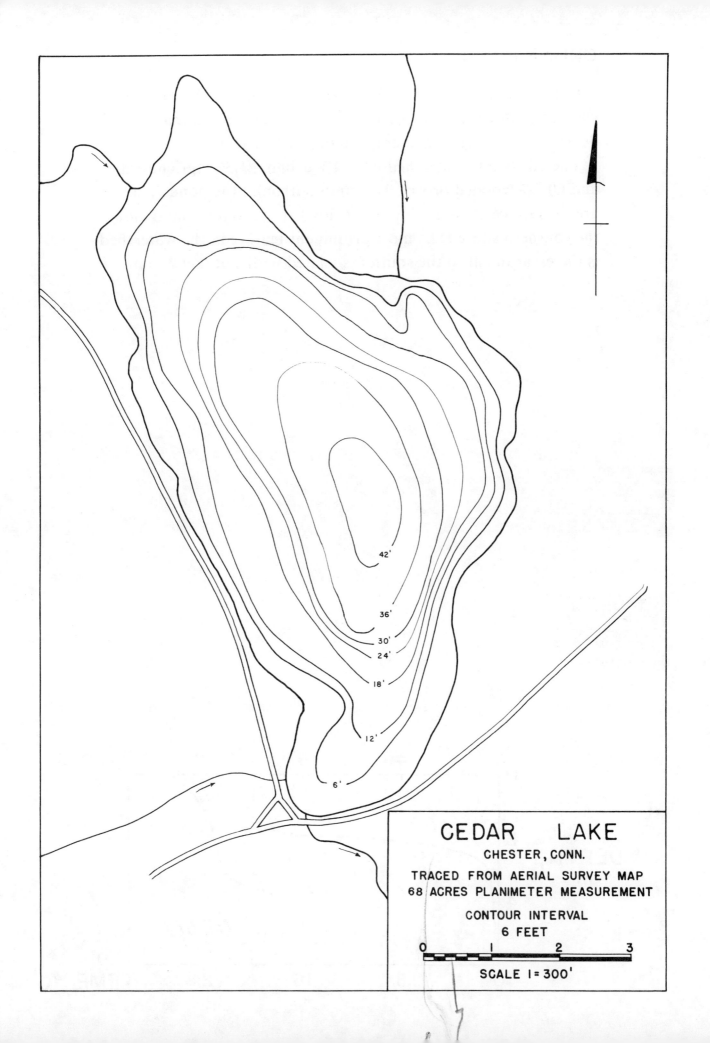

CEDAR POND

Cedar Pond is located in New Haven County in the town of North Branford. It has a surface area of 21.8 acres, a maximum depth of 17 feet and an average depth of 10.9 feet.

The 281.6-acre watershed is 18.4% urban, 21.3% agricultural and 60.3% wooded or wet. Wetlands surround the pond. There are no surface streams. An outlet flows through a wetland along the southern shore. The most prominent feature in the watershed is the escarpment to the south associated with Totoket Mountain.

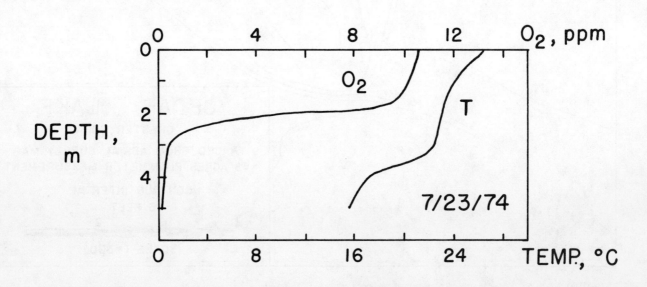

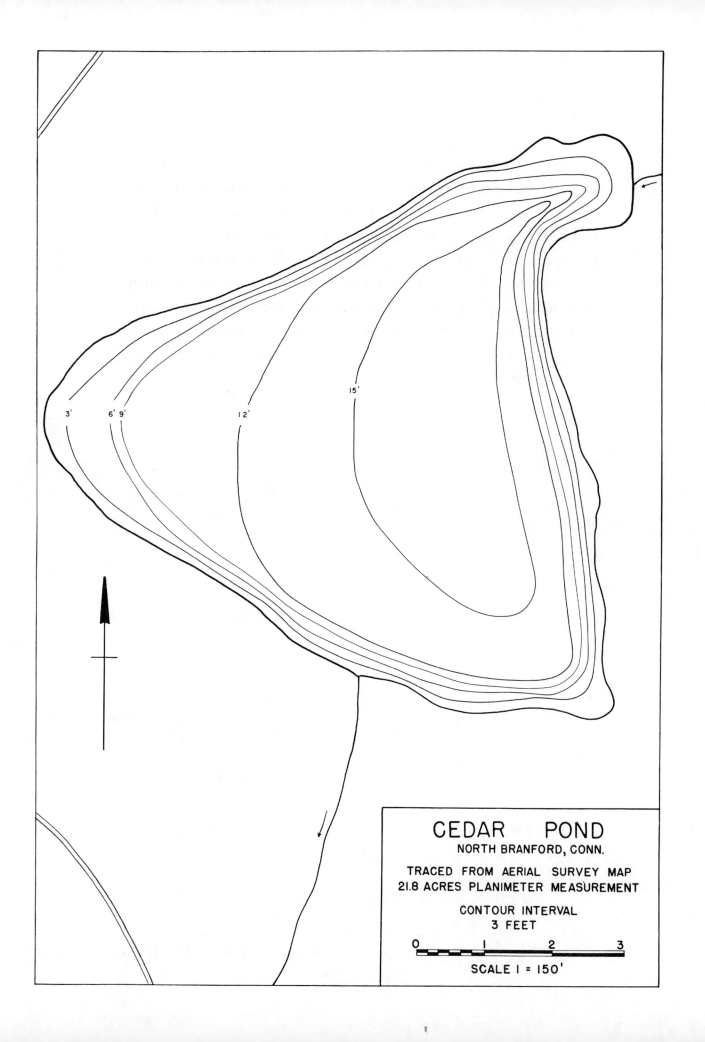

COLUMBIA LAKE

Columbia Lake is located in Tolland County in the town of Columbia. It has a surface area of 277.2 acres, a maximum depth of 26 feet and an average depth of 16.7 feet.

The 1,946-acre watershed is 10.9% urban, 7.4% agricultural and 81.7% wooded or wet. The lake is fed by a wetland and a stream that enter on the southern shore. Columbia Lake Brook drains from the dam that impounds the lake. The surrounding terrain is characterized by moderate slopes of several hillocks.

On August 5, 1980, rooted aquatic weeds were scarce, with only occassional stands of bur reed and pondweeds. Annual winter drawdown may be responsible for this low weed density.

Access for town residents is through a boat launching area and town beach.

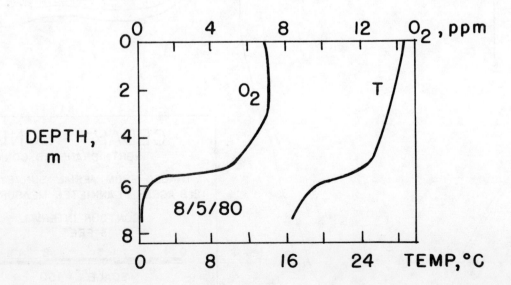

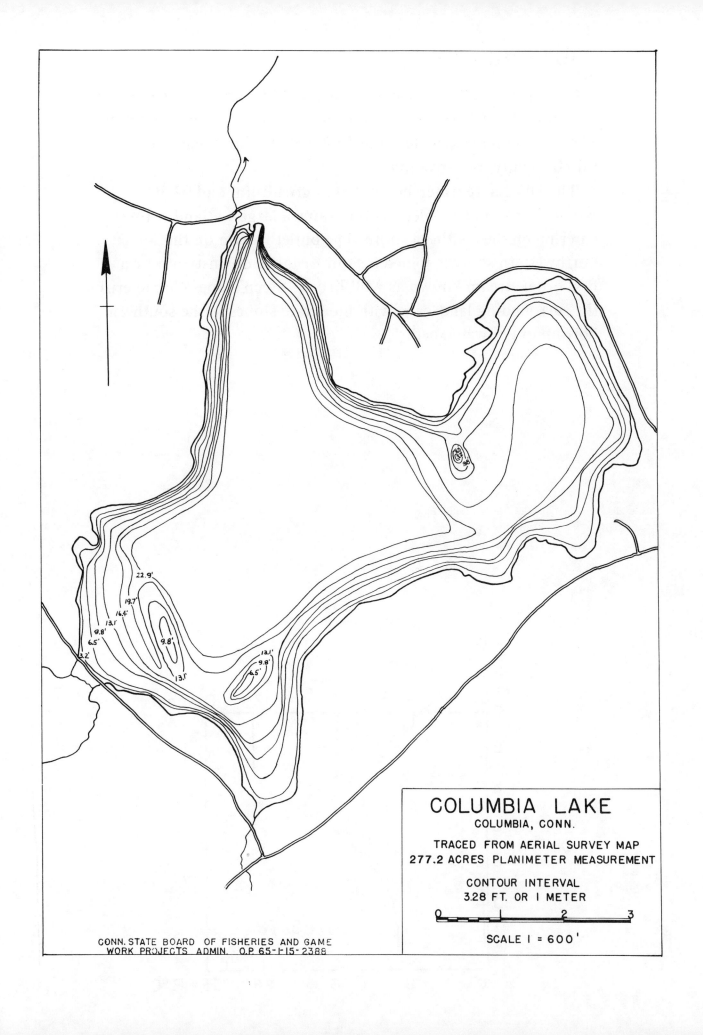

CREAM HILL POND

Cream Hill Pond is located in Litchfield County in the town of Cornwall. It has a surface area of 72 acres, a maximum depth of 43 feet and an average depth of 15.7 feet. The level has been raised slightly by a low dam.

The 403.2-acre watershed is 2.7% agricultural and 97.3% wooded or wet. The inlet stream drains a large wetland before entering on the southern shore. The outlet stream on the northwestern shore is joined a short distance downstream by a first order stream known as Mill Brook. Steep slopes characterize the surrounding landscape with moderate slopes in the southern portion of the watershed.

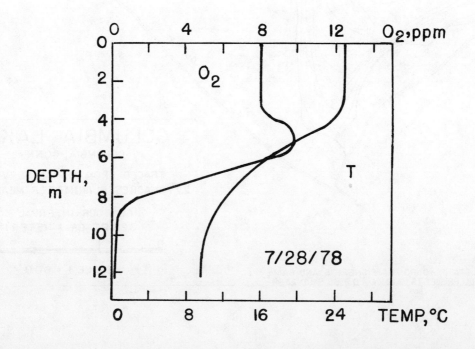

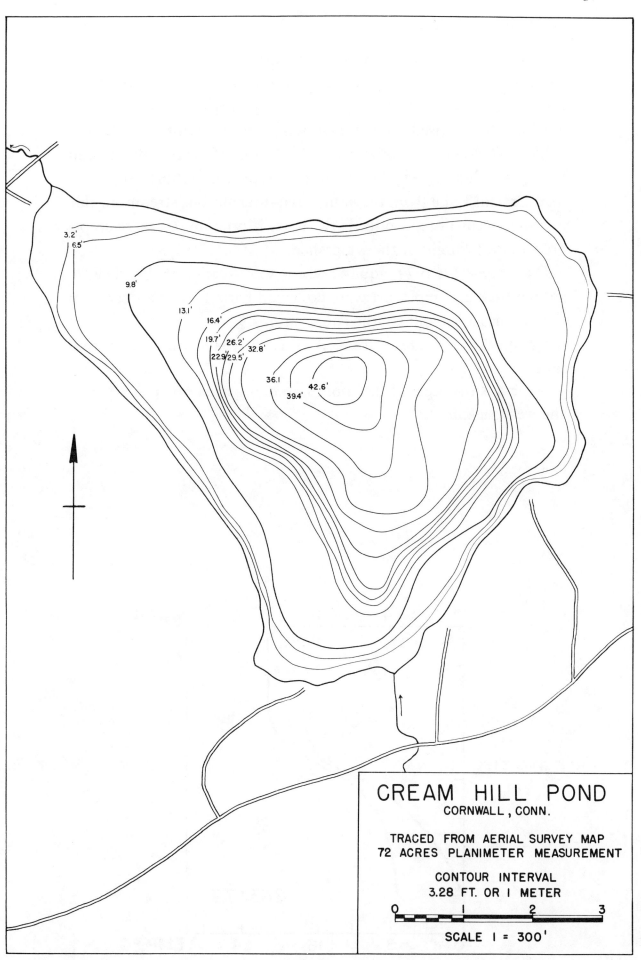

CRYSTAL LAKE

Crystal Lake is located in Tolland County in the towns of Ellington and Stafford. It has a surface area of 200.9 acres, a maximum depth of 50 feet and an average depth of 19.6 feet. The level has been raised approximately six feet by a concrete dam.

The 1,779-acre watershed is 9.7% urban, 2.6% agricultural and 87.7% wooded or wet. An unnamed stream and Aborn Brook feed Crystal Lake from the south. An intermittent stream on the northwestern shore also feeds the lake. Moderate to severe slopes are found throughout the watershed.

On August 13, 1979, aquatic weeds were scarce. Pickerelweed, spatterdock and pipewort were collected from two small coves on the northeastern and southeastern shores.

Crystal Lake has been stocked in recent years with brown and rainbow trout.

Access is through a State boat launching area off Rte. 30 on the western side.

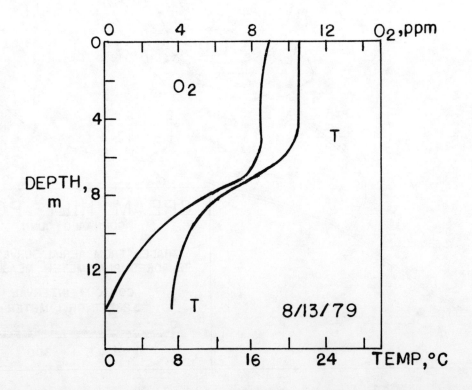

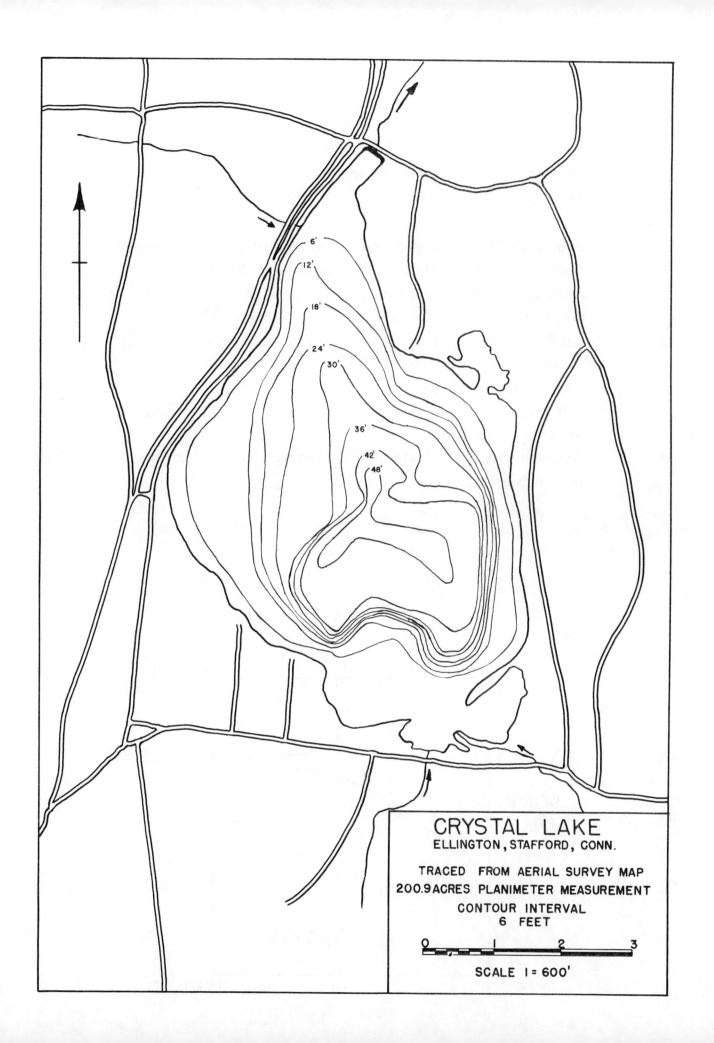

DODGE POND

Dodge Pond is located in New London County in the town of East Lyme. It has a surface area of 33 acres, a maximum depth of 48 feet and an average depth of 20 feet. A dam impounds an additional 7 acres of shallow swamp.

The 428.8-acre watershed is 38.9% urban and 61.1% wooded or wet. The watershed is oriented along a north-south axis. A stream feeds Little Dodge Pond before entering Dodge Pond at the northern end. Topographic variation and slopes in the surrounding area range from moderate to slight.

On August 7, 1979, aquatic weeds were sparse, with some spatterdock present in shallow coves. The shallow area beyond the dam is choked with weeds but should be considered a separate impoundment.

The pond has been stocked with brook, brown and rainbow trout.

Access is through a State boat launching area outside the village of Niantic.

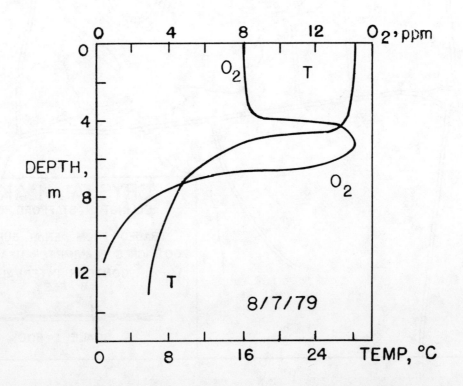

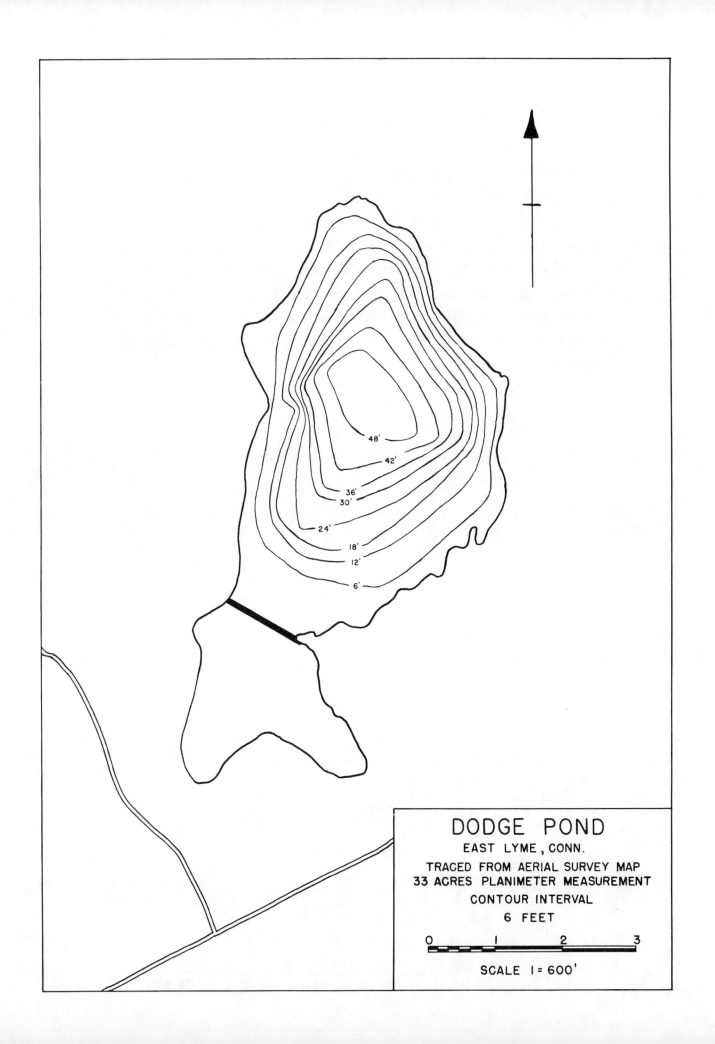

EAGLEVILLE LAKE

Eagleville Lake is located in Tolland County in the towns of Coventry and Mansfield. It has a surface area of 80 acres, a maximum depth of 10 feet and an average depth of 3 feet. It is artificial, having been impounded by an earthen and masonry dam across the Willimantic River.

The watershed contains 71,040 acres. No land use inventory exists.

On August 6, 1980, aquatic weeds were abundant and widespread. Weed beds were particularly dense along the eastern shore. Bigleaf pondweed, pickerelweed, and lesser duckweed were very common with spatterdock, bladderwort and coontail also present.

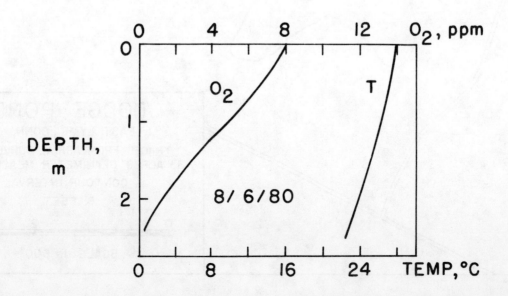

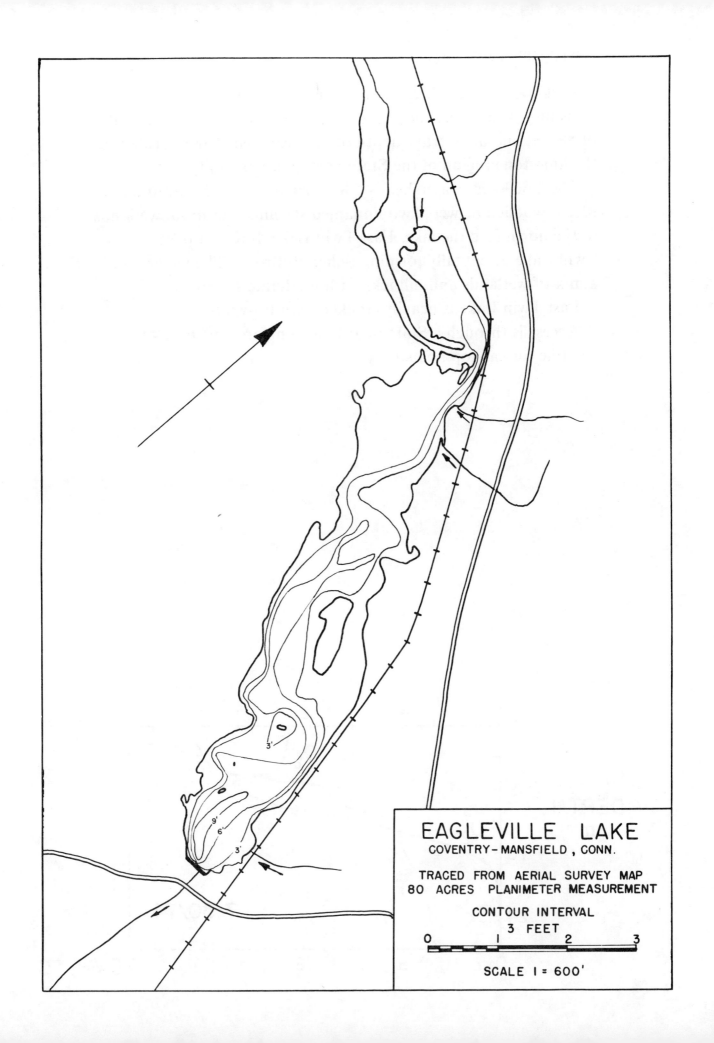

EAST TWIN LAKE

East Twin Lake is located in Litchfield County in the town of Salisbury. It has a surface area of 562.3 acres, a maximum depth of 80 feet and an average depth of 32.4 feet. This lake, located in the limestone region of the State, is high in alkalinity.

The 2,669-acre watershed is 5.8% urban, 3.0% agricultural and 91.2% wooded or wet. Two unnamed streams that drain wetlands to the north feed the lake. East Twin Lake drains into West Twin Lake, eventually forming Schenob Brook. The watershed is a mix of wetlands and hillocks with moderate slopes.

East Twin Lake is heavily stocked with brown trout.

Access is through a State boat launching area off Rte. 44 suitable for car-top boats only.

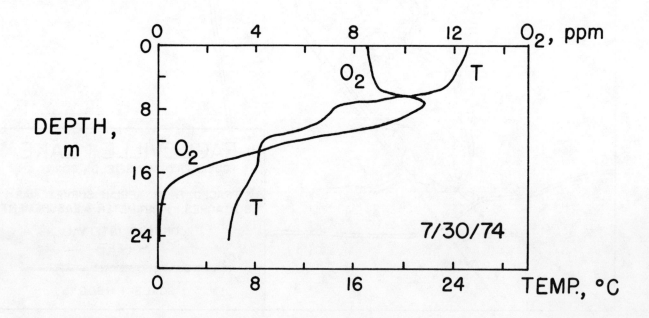

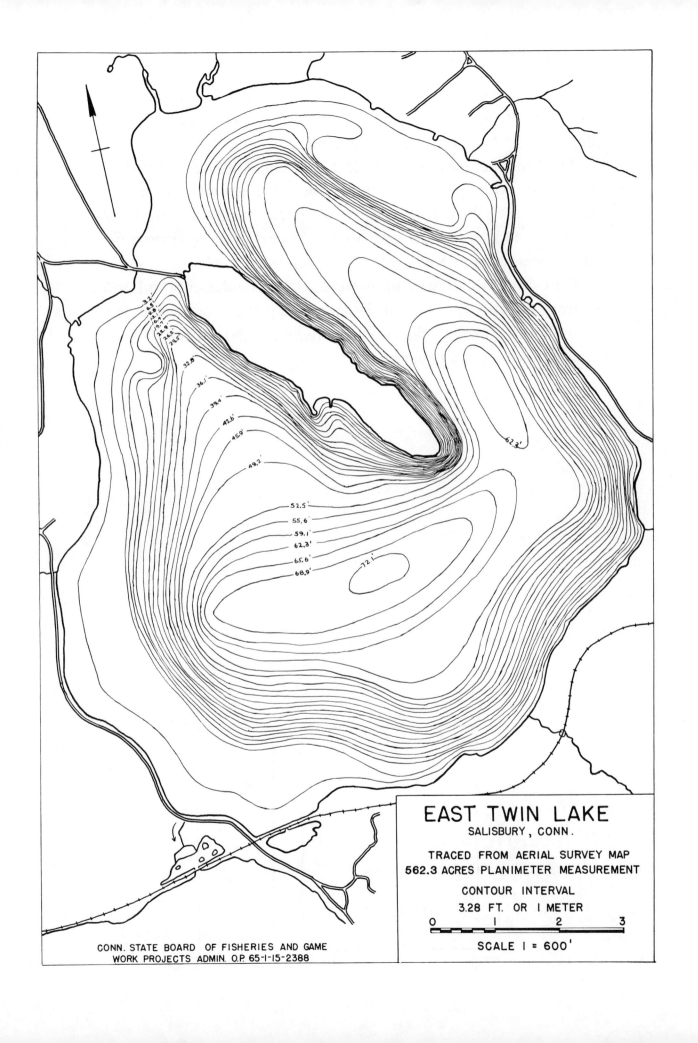

1860 RESERVOIR

The 1860 Reservoir is located in Hartford County in the town of Wethersfield. It has a surface area of 35 acres, a maximum depth of 3.5 feet and an average depth of 2.0 feet.

The 518.4-acre watershed is 45.5% urban, 9.6% agricultural and 44.9% wooded or wet. This impoundment is surrounded almost entirely by contiguous wetlands that drain the watershed in the absence of any defined streams. Relief in the surrounding landscape is slight. No bathymetric map is available.

On October 1, 1979, aquatic weeds, in particular coontail, were extremely dense throughout. During the spring, weeds were dense and about half the surface was covered with mats of the filamentous alga, *Spirogyra*.

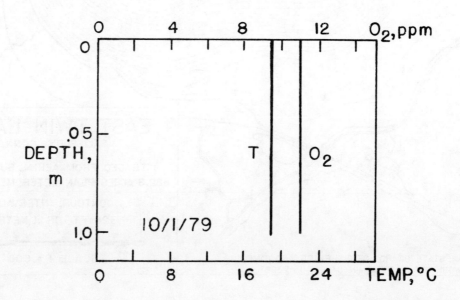

NO BATHYMETRIC MAP AVAILABLE

GARDNER LAKE

Gardner Lake is located in New London County in the towns of Salem, Montville, and Bozrah. It has a surface area of 486.8 acres, a maximum depth of 43 feet and an average depth of 13.7 feet. The level has been raised several feet by a small dam.

The 3,430-acre watershed is 4.0% urban, 11.4% agricultural and 84.6% wooded or wet. Gardner Lake is fed by Sucker Brook, Whittle Brook and several unnamed streams. The outlet is along the northeastern shore through Gardner Brook. The landscape is composed of gentle to moderately sloping terrain with a more irregular, undulating form in the southern portion. A large wetland is contiguous with the northern shore.

On September 9, 1982, a dense weedbed consisting of a mixed growth of primarily stonewort and waterweed with various emergents was noted near the southern inlet. Watershield, white water lily, wild celery and bladderwort formed a second dense bed at the northeastern outlet. Stonewort, spikerush and wild celery were also growing in moderately dense beds along most of the shore.

Gardner Lake has been stocked in recent years with brown and rainbow trout.

Access is through a State boat launching area at the southern end near Rte. 82.

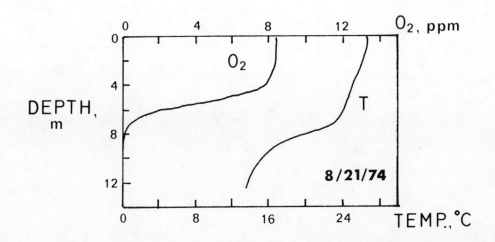

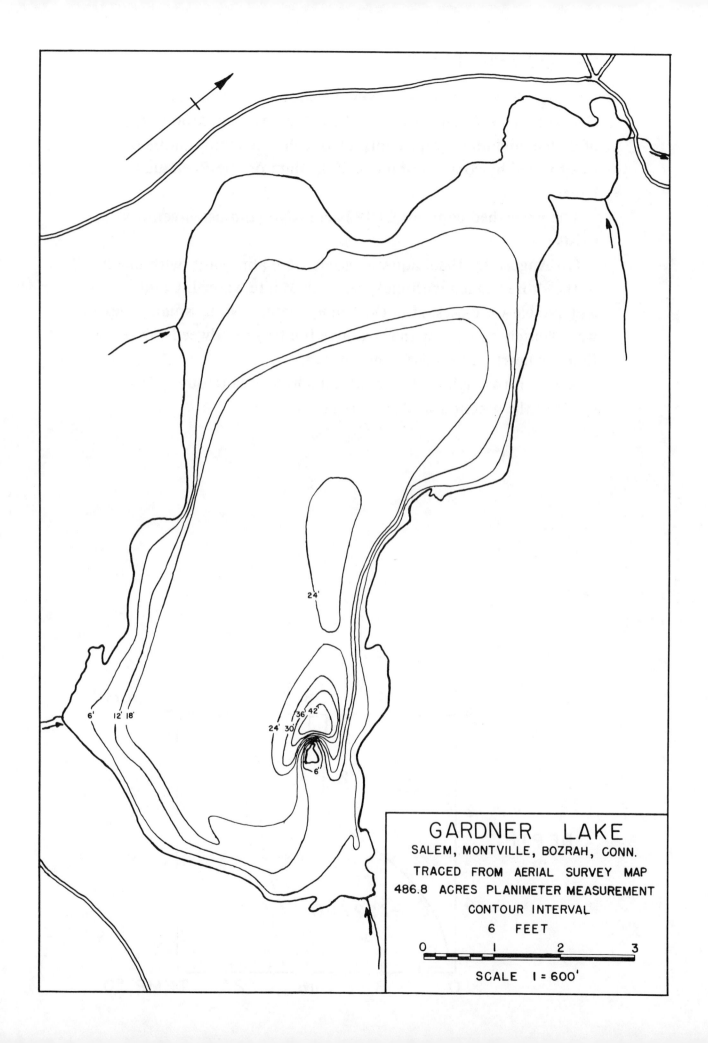

GLASGO POND

Glasgo Pond is located in New London County in the town of Griswold. It has a surface area of 184.2 acres, a maximum depth of 25 feet and an average depth of 10 feet. This impoundment was formed by construction of a large dam on the Pachaug River.

The watershed contains 24,192 acres. No land use inventory exists.

On August 22, 1980, aquatic weeds were abundant, with much of the shore fringed with pickerelweed. White water lily and watershield were present in the deeper waters. Many small coves were densely covered with the above floating pondweeds, little floating heart and snailseed pondweed.

Access is through a State boat launching area about ½ mile north of the junction of Rtes. 201 and 165.

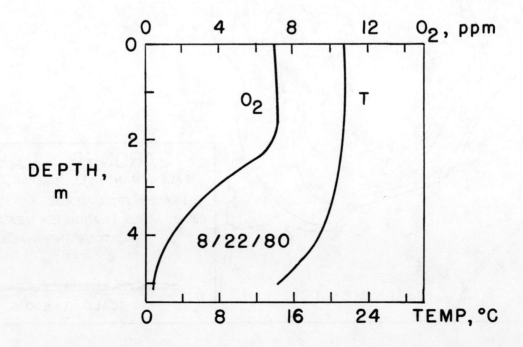

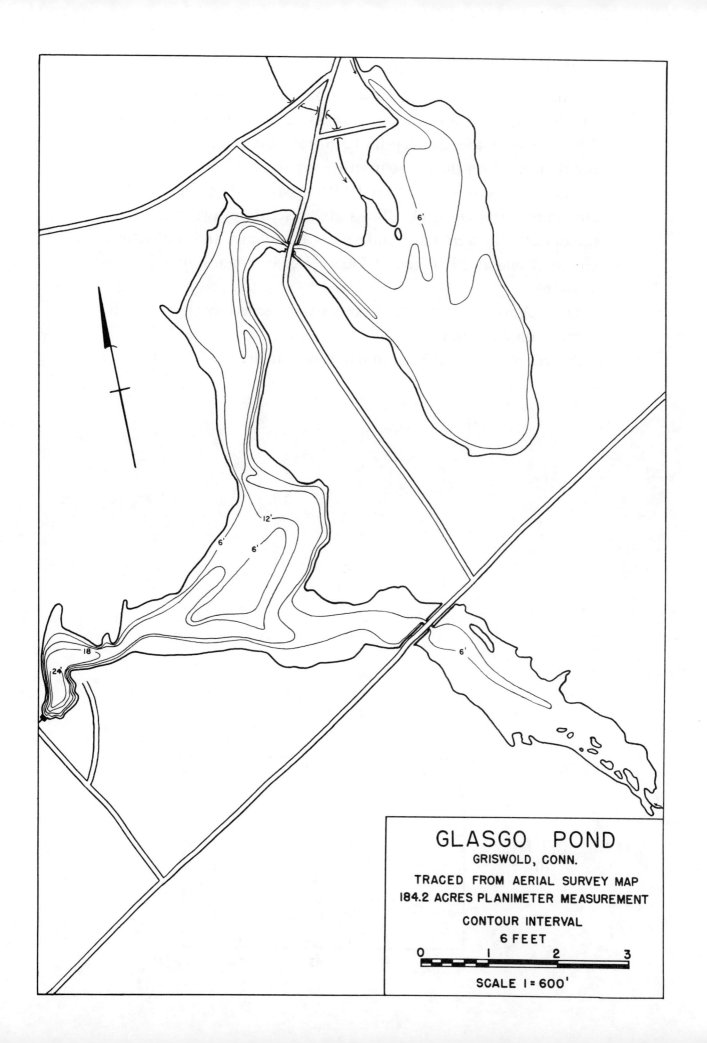

GORTON POND

Gorton Pond is located in New London County in the town of East Lyme. It has a surface area of 53 acres, a maximum depth of 7.5 feet and an average depth of 3.6 feet. It is an artificial impoundment on the Pataganset River.

The 4,154-acre watershed is 18.3% urban, 6.1% agricultural and 75.6% wooded or wet. The Pataganset River originates from the outlet of Powers Lake and serves as the inlet and outlet for Gorton Pond and Pataganset Lake. No bathymetric map is available.

On August 7, 1979, aquatic weeds were generally sparse, with some coontail present.

Access is through a State boat launching area off Rte. 161.

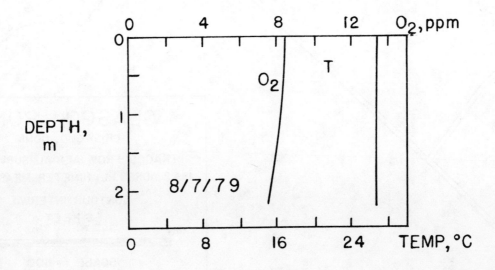

NO BATHYMETRIC MAP AVAILABLE

HAYWARD LAKE

Hayward Lake, also known as Shaw Lake, is located in Middlesex County in the town of East Haddam. It has a surface area of 198.9 acres, a maximum depth of 37 feet and an average depth of 10 feet. The level has been raised slightly by a dam at the southern end.

The 1,592-acre watershed is 6.7% urban, 5.5% agricultural and 87.8% wooded or wet. A number of unnamed streams and wetlands along the northern and eastern shore feed the lake. Lake Hayward Brook is the outlet. The general character of the watershed is moderate slopes with the most severe slopes occurring along the western shore.

Access is through a State boat launching area at the northern end off Rte. 85 south of Colchester.

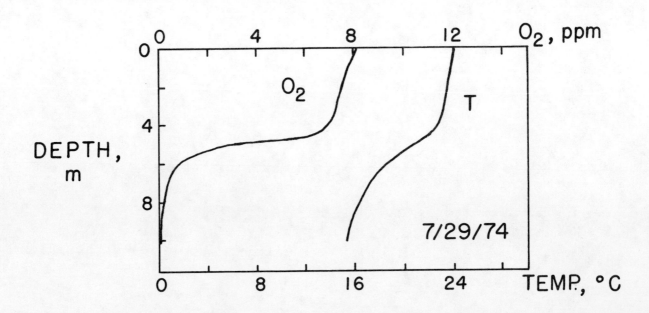

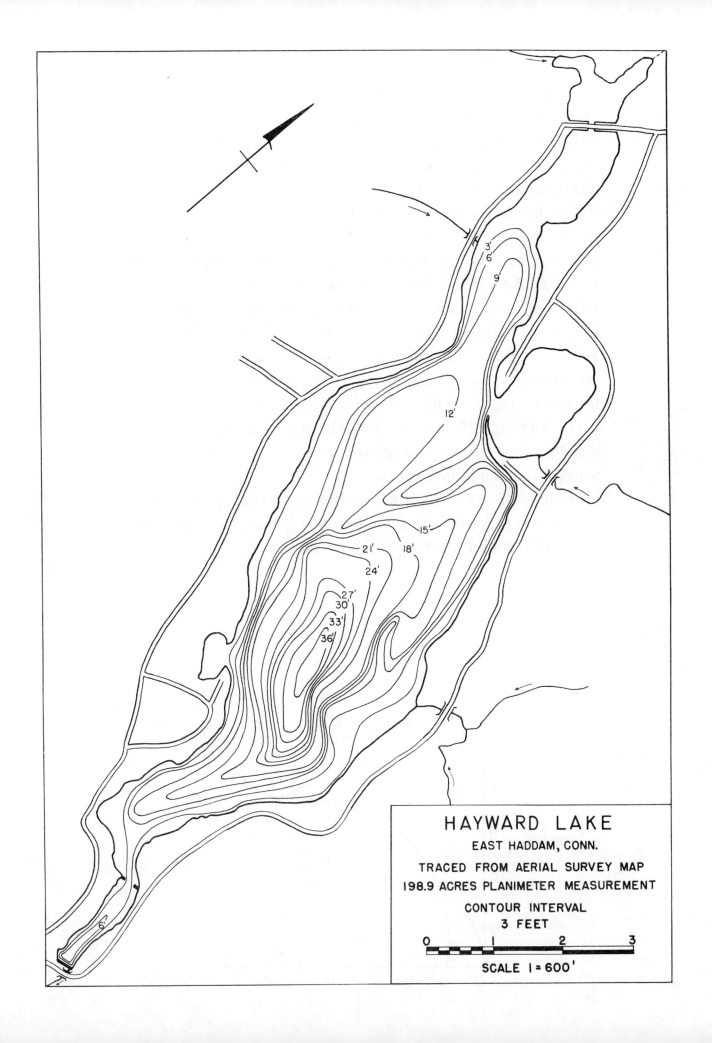

HIGHLAND LAKE

Highland Lake is located in Litchfield County in the town of Winchester. It has a surface area of 444 acres, a maximum depth of 62 feet and an average depth of 19.7 feet. It has three basins; the middle is the deepest. The level has been raised approximately 10 feet by a dam.

The 4,512-acre watershed is 11.2% urban, 2.9% agricultural and 85.9% wooded or wet. Highland Lake lies in a high valley overlooking Winsted. Winchester Club Pond and Crystal Lake feed the lake through Sucker Brook. The outlet of this brook to Highland Lake is controlled by Sucker Brook Dam, an Army Corps of Engineers flood-control project. Taylor Brook is another primary tributary entering on the southern shore. The lake outlet is through a spillway on the northern shore. The terrain is rugged, with a number of severe slopes.

On August 2, 1979, aquatic weeds were relatively sparse. Only a few shallow coves along the shore contained heavy growths of water milfoil and pondweeds.

Highland Lake is stocked with brown and rainbow trout.

Access is through a State boat launching area approximately 100 yards west of the dam at the northern end.

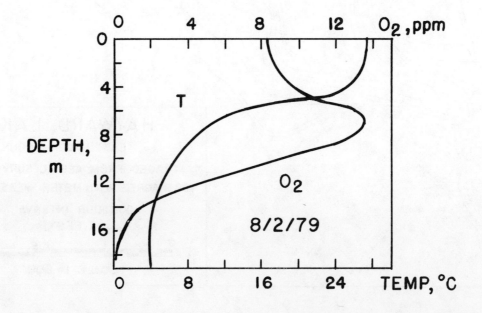

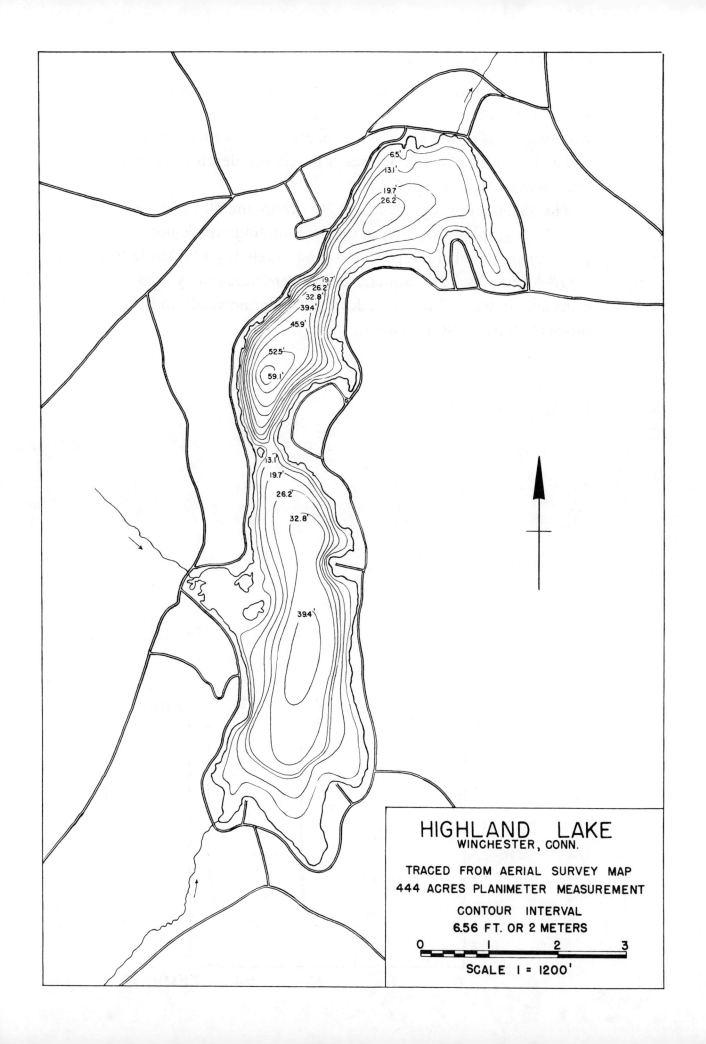

HITCHCOCK LAKES

Hitchcock Lakes are located in New Haven County in the town of Wolcott. Actually a single lake divided by a road and causeway, it is artificial, with a dam along the western shore. It has a surface area of 118.4 acres, a maximum depth of 10 feet and an average depth of 6.2 feet.

The 358.4-acre watershed is 59.2% urban and 40.8% wooded or wet. Topographic variation in the surrounding area is not pronounced; the landscape slopes moderately towards the lake.

On August 28, 1979, aquatic weeds were moderately dense, especially in water 2 to 8 feet deep. Bushy pondweeds and pipewort were most abundant.

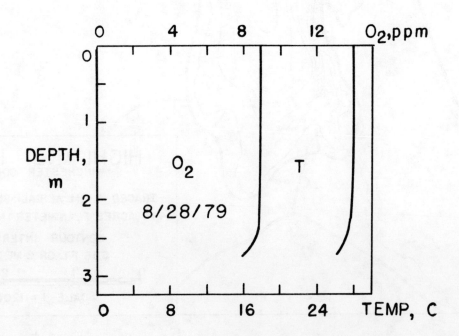

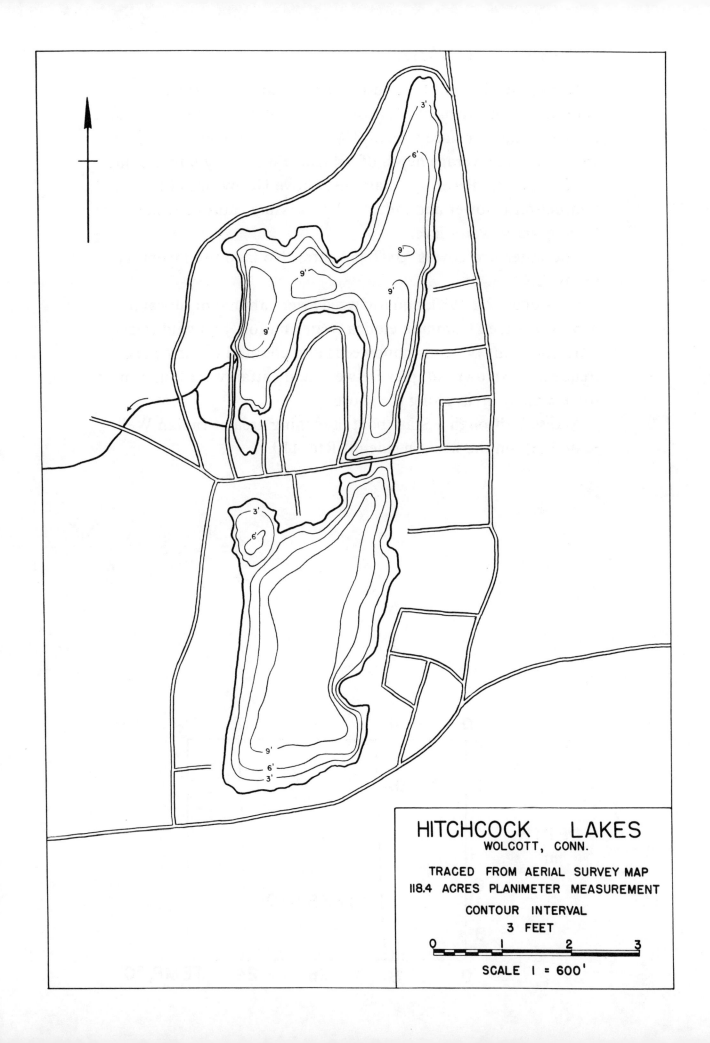

HOUSATONIC LAKE

Housatonic Lake is located in Fairfield and New Haven counties in the towns of Monroe, Shelton, Oxford, Seymour, and Derby. It has a surface area of 328.2 acres, a maximum depth of 26 feet and an average depth of 9.4 feet. Formed by impounding the Housatonic River by a dam just above Derby, water use for hydroelectric power and industrial processing causes considerable fluctuation in water level.

The watershed covers 1,007,360 acres. No land use inventory exists.

On August 28, 1980, aquatic weeds were abundant in water depths to 9 feet. Common elodea, water milfoil, and wild celery were abundant. Various pondweeds and coontail occurred less frequently. Arrowhead, pickerelweed and cattail were found in some marshy areas along the shore.

Access is through a State boat launching area at Indian Well State Park on the western side off Rte. 110.

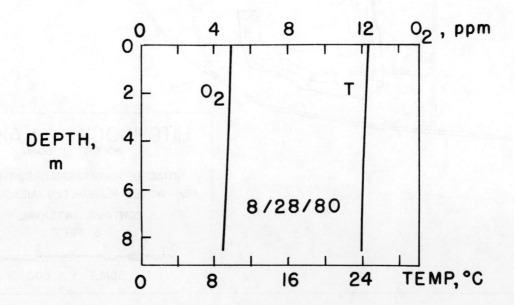

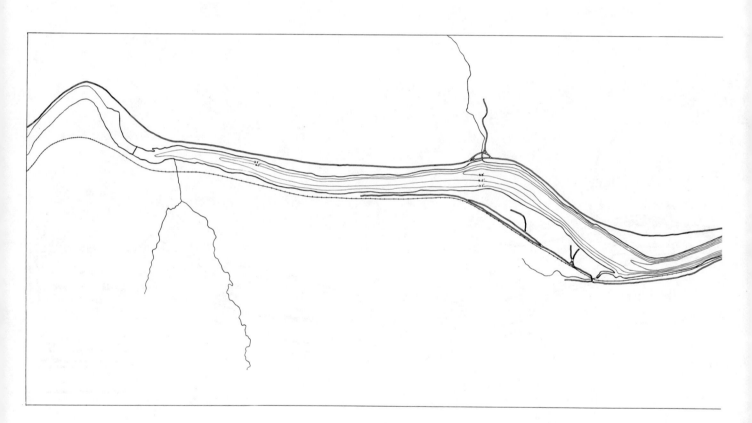

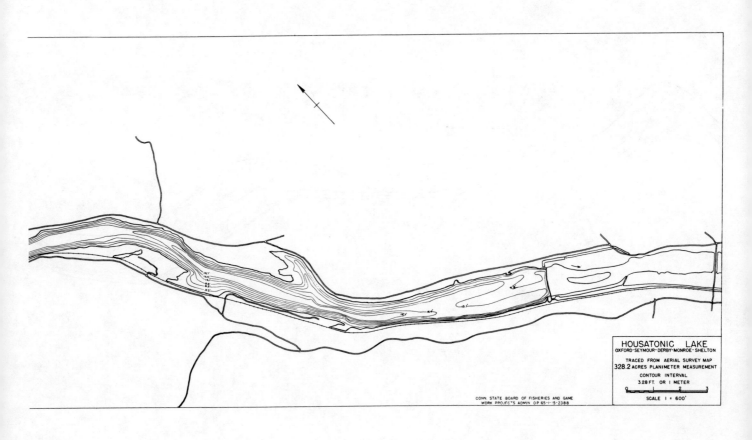

KENOSIA LAKE

Kenosia Lake is located in Fairfield County in the town of Danbury. It has a surface area of 56 acres, a maximum depth of 15 feet and an average depth of 11.6 feet.

The 3,149-acre watershed is 27.4% urban, 1.6% agricultural and 71.0% wooded or wet. Relatively large wetland areas are east and northwest of the lake. Slight to moderate slopes prevail throughout the watershed. No bathymetric map is available.

On July 31, 1980, aquatic weeds were dense and widespread. Thick beds of water milfoil covered much of the bottom to depths of 9 feet. White water lily and spatterdock coated the surface waters along the northern and southern ends, with patches of pickerelweed found all along the shore.

Access is through a State boat launching area off Rte. 202. Lake Kenosia Town Park offers access to town residents.

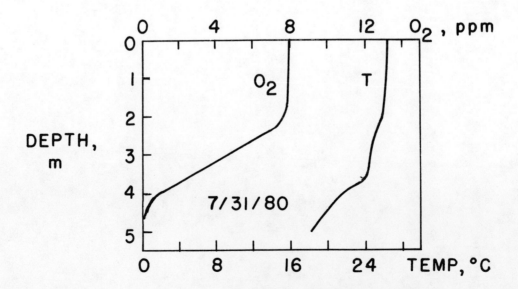

NO BATHYMETRIC MAP AVAILABLE

LILLINONAH LAKE

Lillinonah Lake is located in Fairfield, New Haven and Litchfield Counties in the towns of Newtown, Brookfield, New Milford, Bridgewater, and Southbury. It has a surface area of 1,900 acres, a maximum depth near the dam of 100 feet and an average depth of 25 feet. It was created in 1955 by impounding the Housatonic River for hydroelectric power generation.

The 895,360-acre watershed is 7.9% urban, 16.5% agricultural and 75.6% wooded or wet.

A phosphorus and water budget for Lake Lillinonah was developed by Aylor and Frink (1980). Phosphorus removal from the Danbury sewage treatment plant was begun in 1977 (Jones and Lee, 1981). This, coupled with phosphorus removal at Pittsfield, MA (Charles Fredette, personal communication) is expected to reduce phosphorus concentrations. Distribution of PCBs in the bottom sediments is described by Frink, et al. (1982).

Access is through a State boat launching area on the northern shore off Rte. 133 about 9 miles south of New Milford. A second access is at the mouth of Pond Brook off Hanover Road approximately 3 miles north of Rte. I-84.

The bathymetric map of Lake Lillinonah is reprinted by permission of the International Map Company.

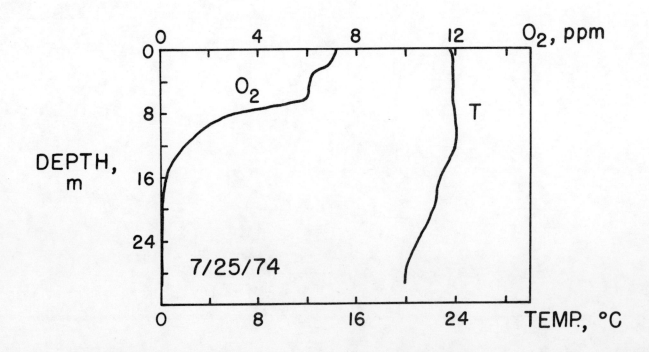

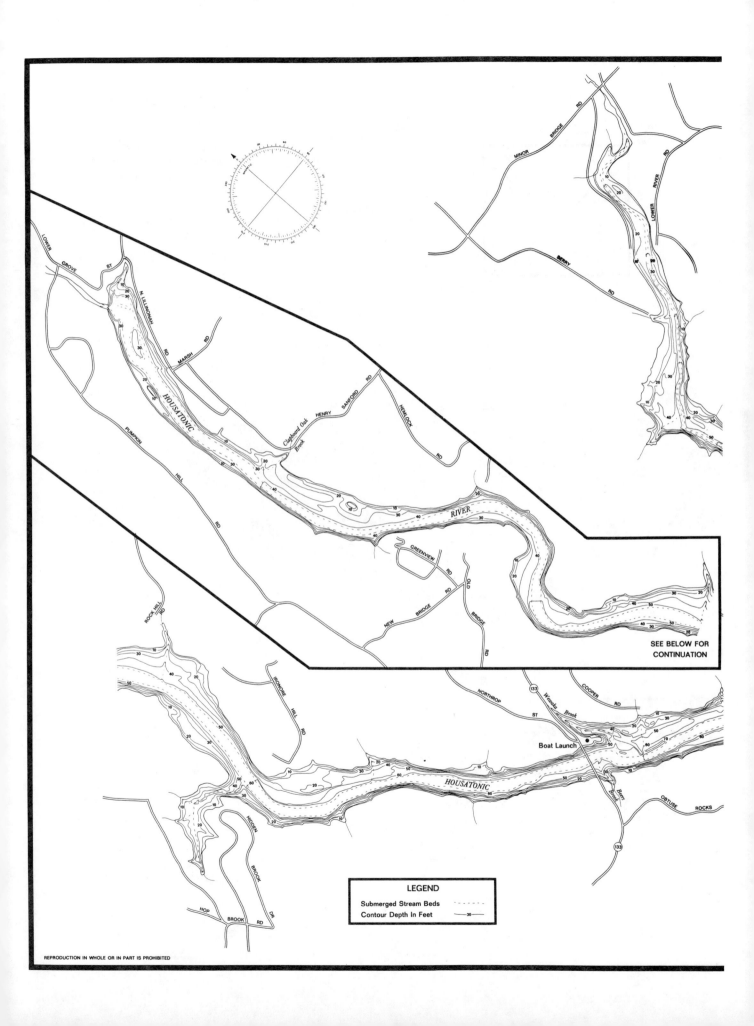

Angler's Aid™
SUB-SURFACE STRUCTURE MAP OF
LILLINONAH LAKE
FAIRFIELD COUNTY - LITCHFIELD COUNTY
CONNECTICUT

NORMAL POOL ELEVATION 198 FEET

DEPTH CONTOURS ARE AT 10 FOOT INTERVALS

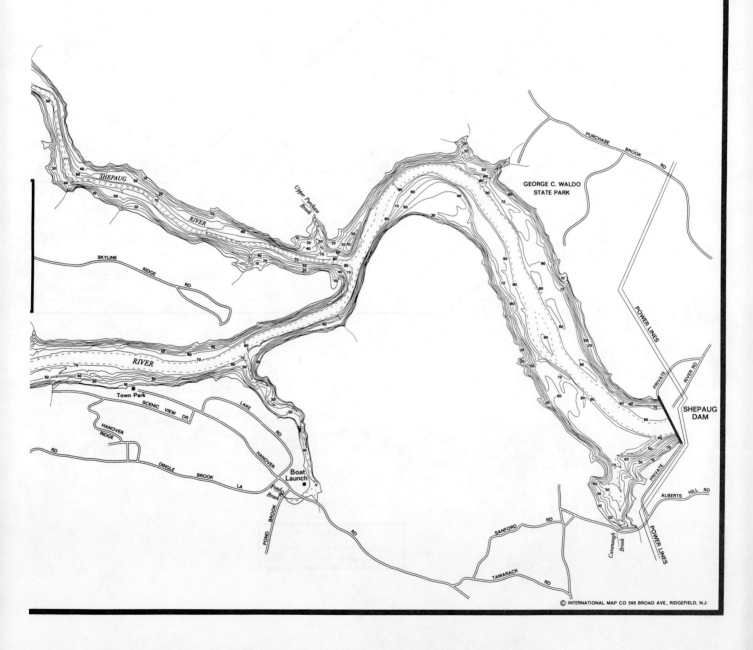

© INTERNATIONAL MAP CO 595 BROAD AVE, RIDGEFIELD, N.J.

LINSLEY POND

Linsley Pond is located in New Haven County in the towns of Branford and North Branford. It has a surface area of 23.3 acres, a maximum depth of 44 feet and an average depth of 20.5 feet.

The 582.4-acre watershed is 31.5% urban, 9.8% agricultural and 58.7% wooded or wet. Linsley Pond is fed primarily from a stream that drains from Cedar Pond to the northeast. The landscape along the eastern shore is relatively flat. Severe slopes occur along portions of the western shore. Pisgah Brook is the outlet.

Linsley Pond has been studied extensively by Hutchinson and associates (Hutchinson, 1957).

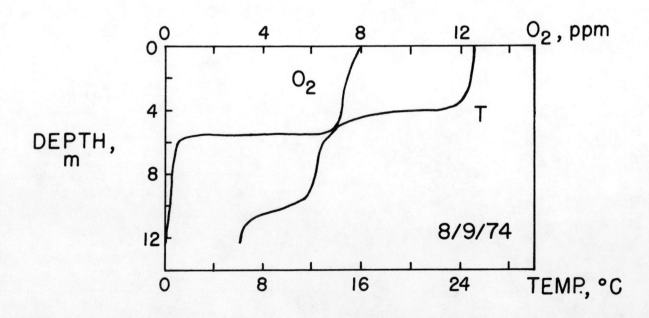

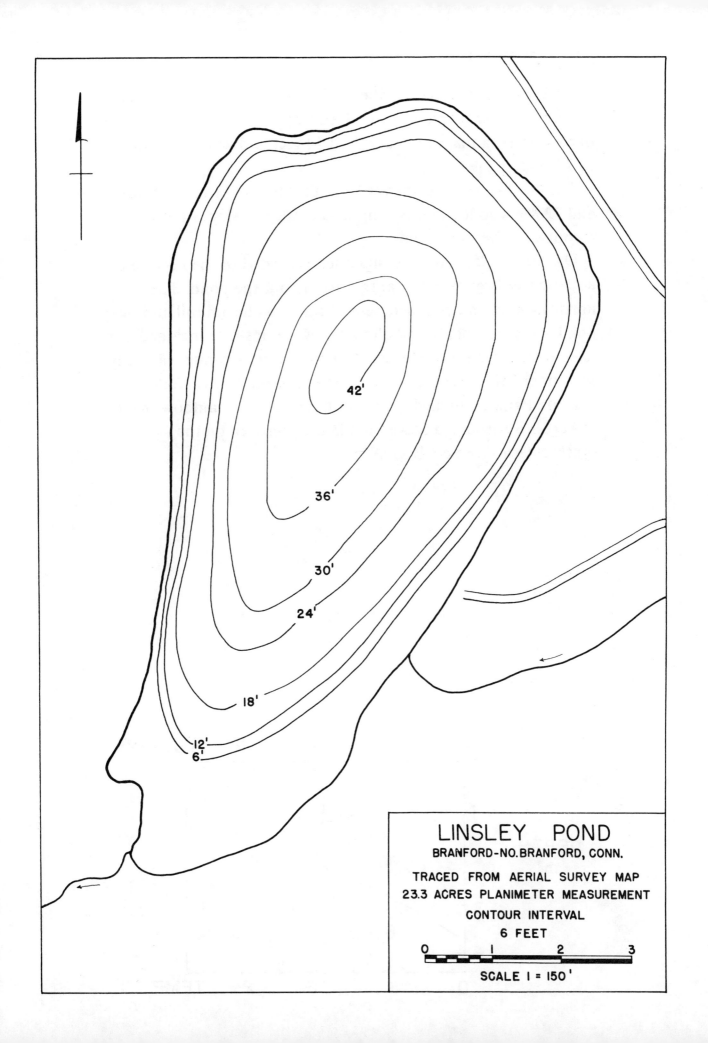

LITTLE POND

Little Pond, sometimes called Schoolhouse Pond, is located in Windham County in the town of Thompson. It has a surface area of 65.4 acres, a maximum depth of 14 feet and an average depth of 7.8 feet.

The 460.8-acre watershed is 8.6% urban, 18.6% agricultural and 72.8% wooded or wet. Slight to moderate slopes occur throughout the watershed.

On August 25, 1980, transparency extended to the bottom and aquatic weeds grew abundantly throughout the pond. The northern portion contained dense beds of water milfoil and bushy pondweeds. Species found along the western shore included bur reed, cattail and pickerelweed. A dense stand of white water lily surrounded the outlet. Bladderwort appeared occassionally.

Little Pond is stocked with brook, brown and rainbow trout.

Access is through a State boat launching area on the southeastern shore off Rte. 193.

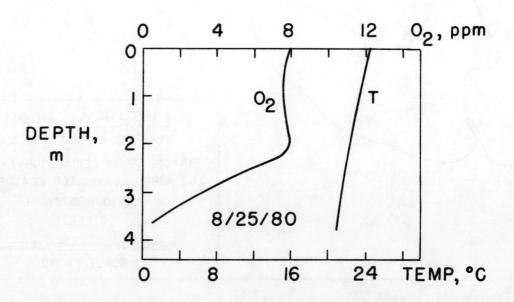

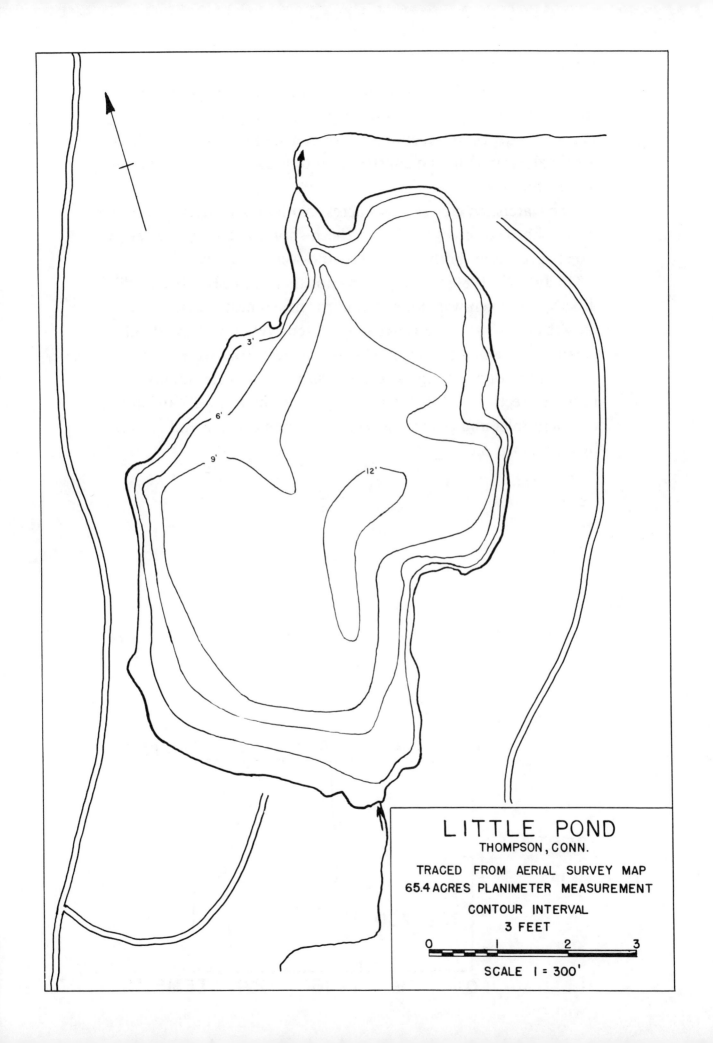

LONG MEADOW POND

Long Meadow Pond is located in Litchfield County in the town of Bethlehem. It has a surface area of 110.5 acres, a maximum depth of 7 feet and an average depth of 4.2 feet. It is artificial, formed by the construction of a dam across the mouth of a small valley.

The watershed covers 1,043 acres. No land use inventory exists. A small brook enters on the northern shore. Moderate to severe slopes prevail on both the eastern and western shores.

On July 28, 1980, the pond was not thermally stratified and dissolved oxygen was present in even the deepest areas. Dense growths of aquatic weeds occurred extensively throughout. The bottom was covered with muskgrass; bigleaf pondweed and wild celery were found along the eastern shore and in the shallow northern section. White water lily grew in thick stands around the northern inlet and covered a section at the northern end cut off by an earthen dam.

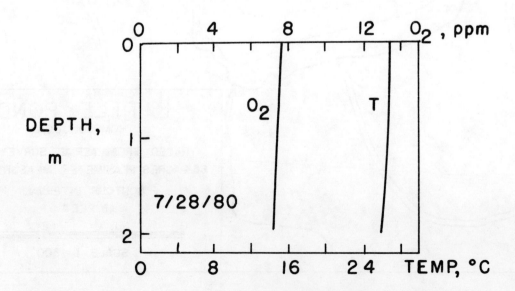

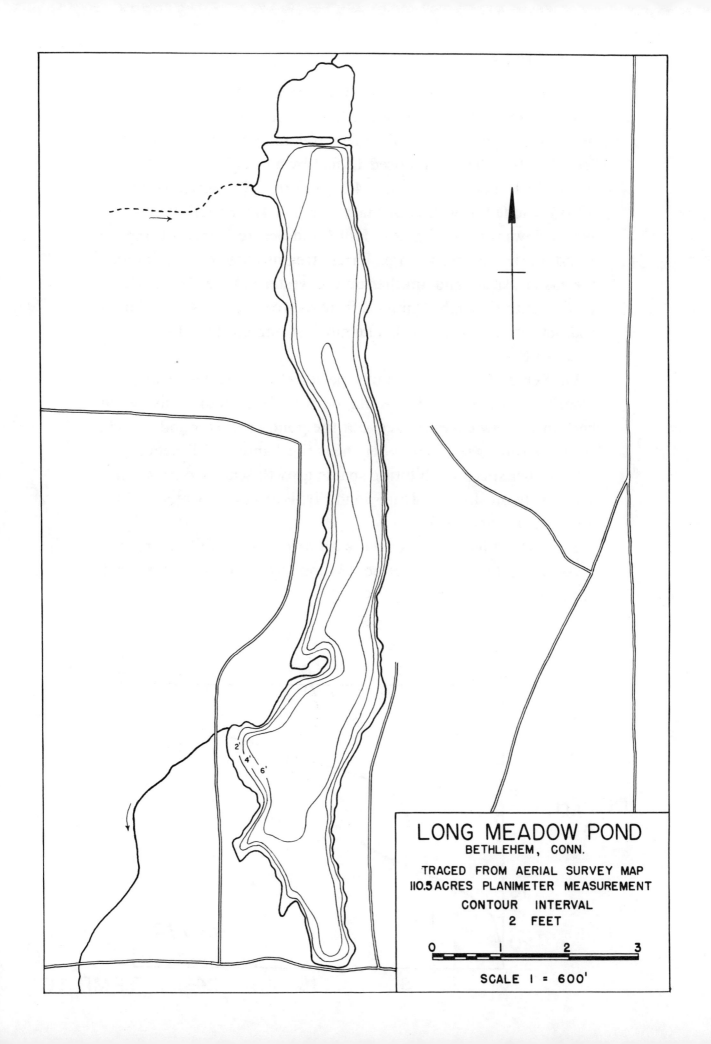

LONG POND

Long Pond is located in New London County in the towns of Ledyard and North Stonington. It has a surface area of 98.6 acres, a maximum depth of 72 feet and an average depth of 15.2 feet. The level has been raised 12 feet by a dam.

The 2,918-acre watershed is 4.0% urban, 3.9% agricultural and 92.1% wooded or wet. A primary inlet is Lantern Hill Brook, which flows through Lantern Hill Pond before it enters Long Pond on the northern shore. Three streams enter the lake along the eastern shore and another on the western shore. The outlet is to the south through Whitford Brook. Moderate to severe slopes characterize the watershed. The most prominent feature is Lantern Hill.

On September 12, 1977, nine species of aquatic weeds were identified. Water milfoil and bladderwort formed extensive weed beds in shallow coves throughout the pond. Yellow pond lily and swamp shrubs grew in numbers near the Lantern Hill Brook inlet. Other species exhibiting sparse growth scattered along the shore were muskgrass, white pond lily, wild celery, watershield, and several pondweeds.

Long Pond has been stocked with brown and rainbow trout.

Access is through a State boat launching area on Lantern Hill Road, off Rte. 214 or Rte. 184.

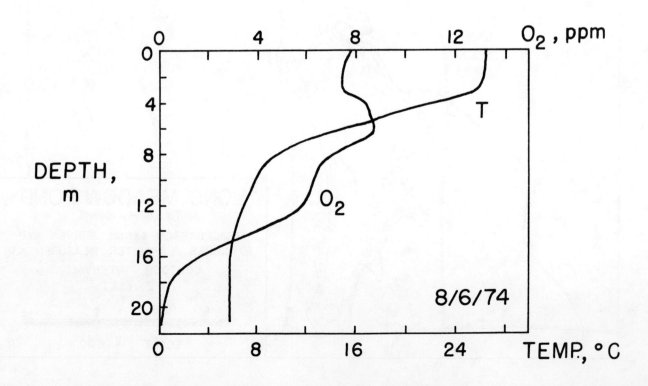

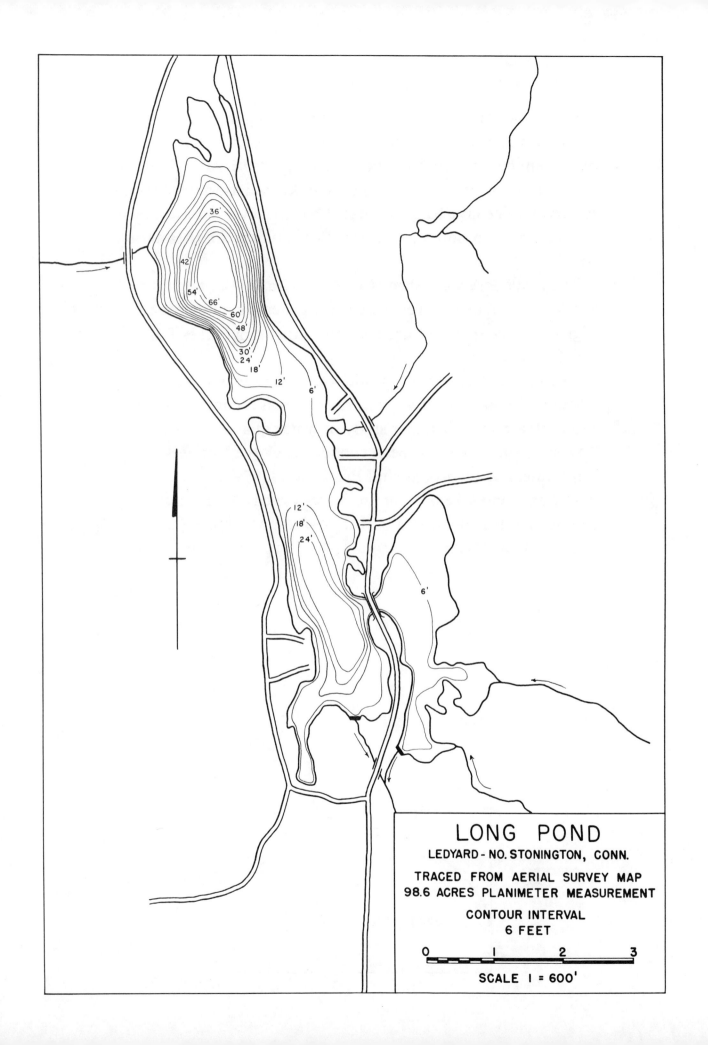

LOWER BOLTON LAKE
(Lower Willimantic Reservoir)

Lower Bolton Lake is located in Tolland County in the towns of Vernon and Bolton. It has a surface area of 178.4 acres, a maximum depth of 26 feet and an average depth of 11.3 feet. It is the lower portion of a three-lake chain known as the Willimantic Reservoir. The lake basin is natural but the level has been raised by a concrete, masonry and earthen dam on the southeastern end.

The 2,419-acre watershed is 10.2% urban, 5.0% agricultural and 84.8% wooded or wet. Slopes are generally moderate to slight. Surface flow and drainage from the upper lakes feed Lower Bolton Lake.

On September 1, 1979, aquatic weeds were scarce, apparently reflecting the success of winter drawdown as a control measure. A diagnostic and feasibility study of water quality in Middle and Lower Bolton Lakes has been conducted by the Connecticut Department of Environmental Protection (1979b).

The dam and lake bottom are State-owned. Access is through a State boat launching area on U.S. 44A, approximately one mile east of Bolton Notch.

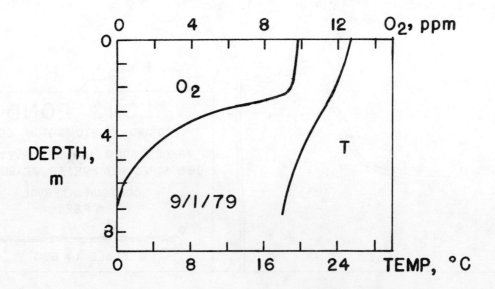

LOWER BOLTON LAKE

BOLTON, VERNON, CONN.

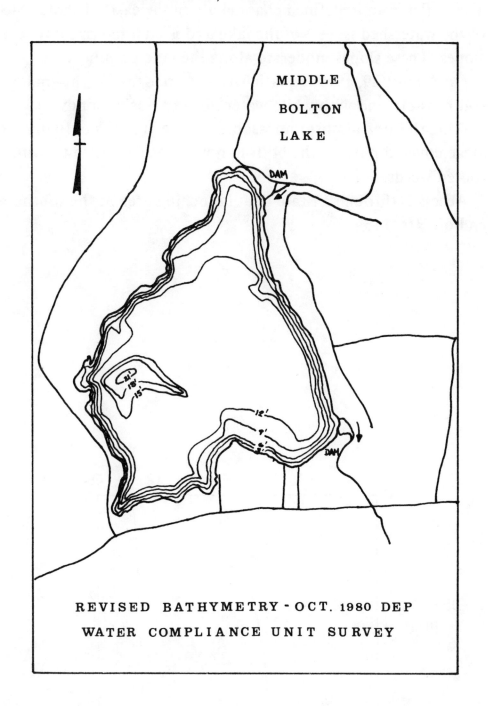

REVISED BATHYMETRY - OCT. 1980 DEP
WATER COMPLIANCE UNIT SURVEY

MAMANASCO LAKE

Mamanasco Lake is located in Fairfield County in the town of Ridgefield. It has a surface area of 95 acres, a maximum depth of 10 feet and an average depth of 6.9 feet. The level has been raised slightly by a small dam.

The 537.6-acre watershed is 66.7% urban and 33.3% wooded or wet. The inlet is a surface flow that originates in Turtle Pond. The outlet is an undefined channel along the eastern shore. Most of the watershed is west of the lake and is characterized by severe slopes. These slopes moderate along the lake periphery.

On August 29, 1979, aquatic weeds were dense in the shallow southeastern end with white water lily and bushy pondweed making the area almost impassable. Dense growths of filamentous algae covered most of the bottom not already overgrown with rooted weeds.

Access is through a State boat launching area at the southern end off Rte. 116.

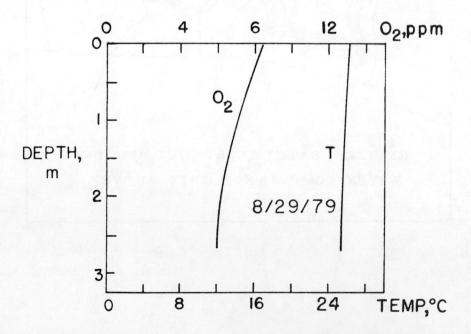

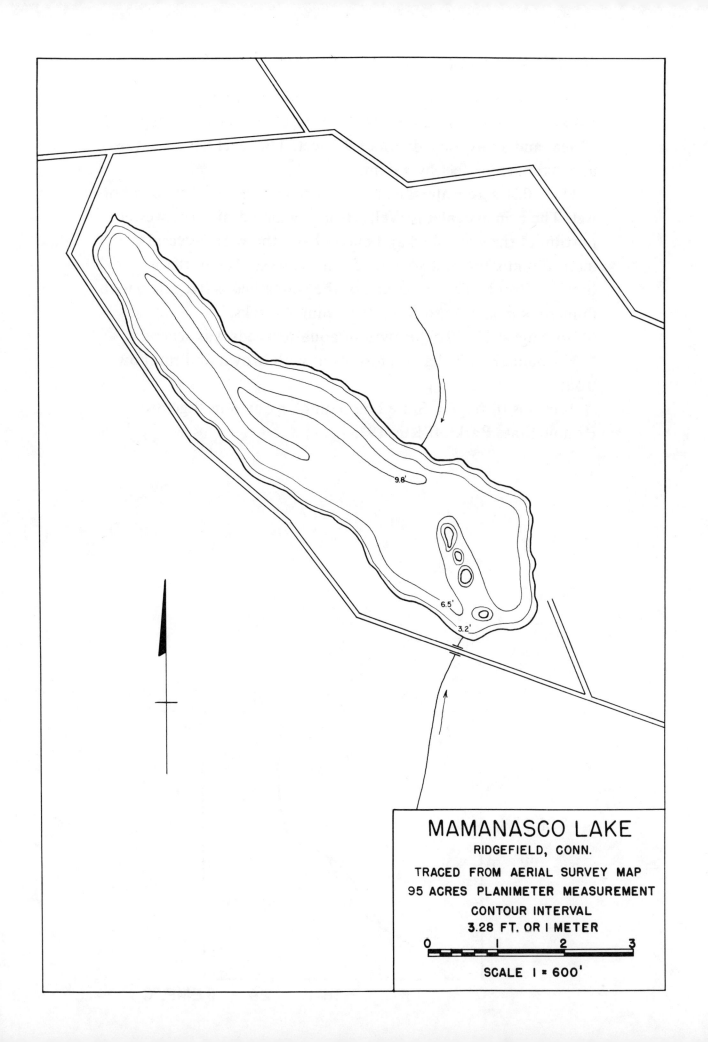

MASHAPAUG LAKE

Mashapaug Lake is located in Tolland County in the town of Union. It has a surface area of 297.1 acres, a maximum depth of 43 feet and an average depth of 9.2 feet. The level has been raised approximately 8 feet by a dam.

The 3,021-acre watershed is 3.3% urban and 96.7% wooded or wet. The primary inlet is Wells Brook, which drains the western portion of the watershed and enters from the west. Several intermittent streams also enter from the west. The outlet is Bigelow Brook. The landscape of the watershed is irregular, with numerous moderate to severely sloping hillocks.

On August 13, 1979, growth of aquatic weeds was scarce.

Mashapaug Lake has been stocked with brown and rainbow trout.

Access is through a State boat launching area in Bigelow Hollow State Park off Rte. 197.

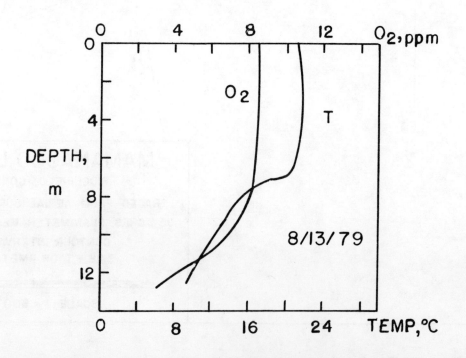

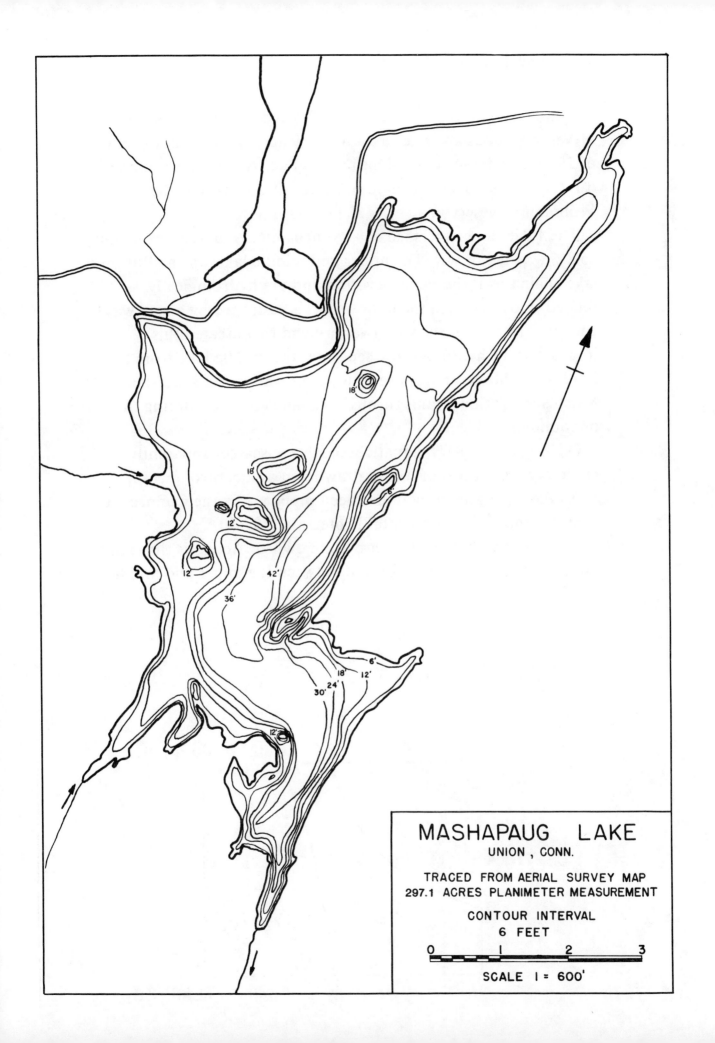

MIDDLE BOLTON LAKE
(Middle Willimantic Reservoir)

Middle Bolton Lake is located in Tolland County in the town of Vernon. It has a surface area of 114.9 acres, a maximum depth of 26 feet, and an average depth of 9.8 feet. As the name implies, it is the middle lake in a chain of three lakes known as the Willimantic Reservoir.

The 1,946-acre watershed is 8.0% urban, 6.2% agricultural, and 85.8% wooded or wet. The upper and middle lakes of the Bolton lake chain are at the same elevation and are hydrologically related. Flow between them, however, is restricted by a roadway culvert. The dam at the southwestern end that created this impoundment hydrologically separates the middle lake from the lower lake. Moderate to severe slopes prevail in the watershed. Much surface flow drains through wetlands before entering this impoundment.

On August 21, 1979, aquatic weeds were scarce, apparently reflecting the success of winter drawdown as a control measure.

A lake management plan has been prepared by the Connecticut Department of Environmental Protection (1979b).

The dam and lake bottom are State owned. Access is through a State boat launching area located between the upper and middle lakes.

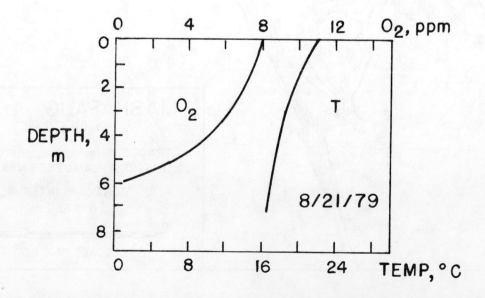

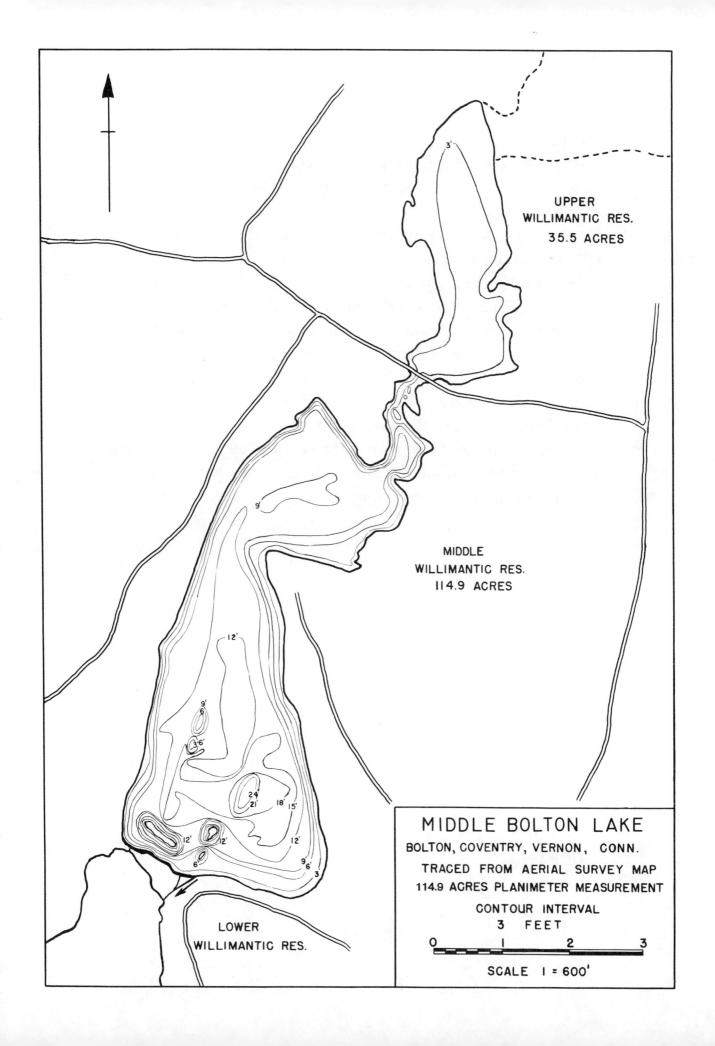

MOODUS RESERVOIR

Moodus Reservoir is located in Middlesex County in the town of East Haddam. It has a surface area of 451 acres and a maximum depth of 10 feet. The mean depth has not been calculated. It is a large, shallow, artificial impoundment consisting of two basins separated by a roadway.

The watershed contains 6,720 acres. No land use inventory exists. Moodus Reservoir is fed by several tributary streams entering on the northern shore.

On July 25, 1980, aquatic weeds were dense in both basins, particularly in areas where the water depth was three feet or less. Emergent weeds in the northern basin included white water lily and pickerelweed. Submerged weeds were mostly fanwort and bladderwort. The southern basin contained an emergent mix of watershield, white water lily and spatterdock, and a submerged flora similar to that in the northern basin.

Access to the southern basin is through a State boat launching area outside Moodus, just south of the causeway between the basins. Access to the northern basin is off Rte. 149.

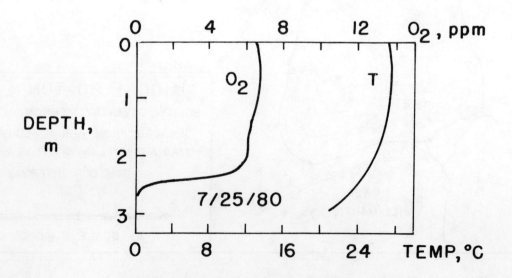

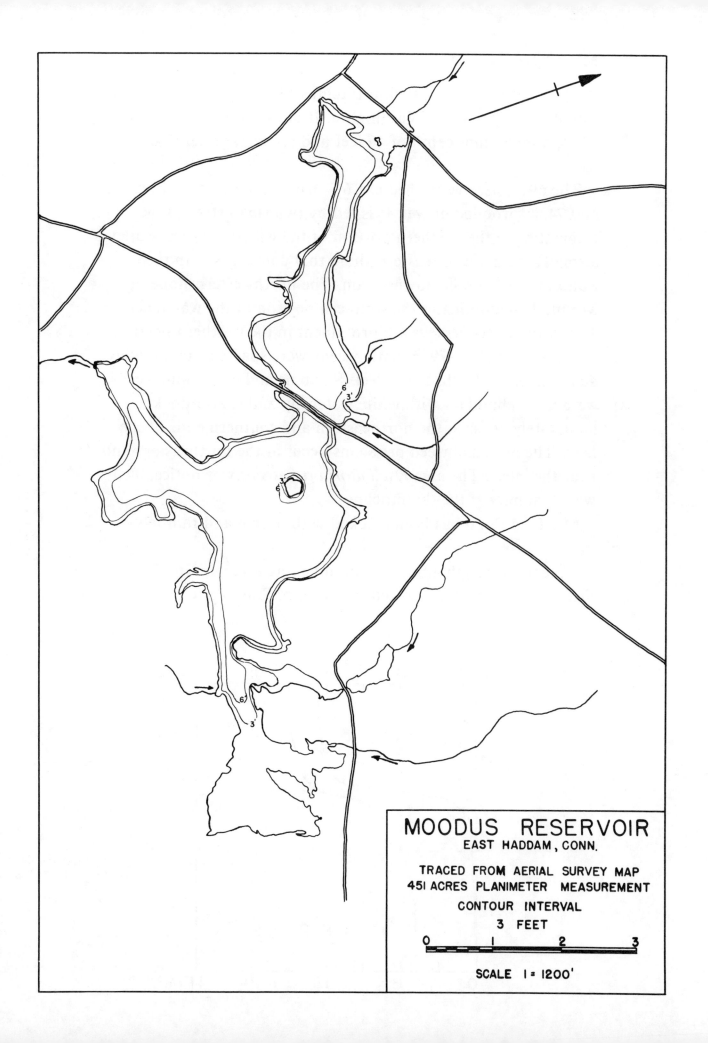

MT. TOM POND

Mt. Tom Pond is located in Litchfield County in the towns of Litchfield, Morris and Washington. It has a surface area of 61.5 acres, a maximum depth of 46 feet and an average depth of 21.2 feet.

The 691.2-acre watershed is 1.6% urban, 24.0% agricultural and 74.4% wooded or wet. It is fed by two small streams originating in the northern portion of the watershed. The outlet is a small stream on the eastern shore that converges with the Bantam River a mile downstream. The northwestern slope of Mount Tom dominates the southern portion of the watershed. Two pronounced valleys are prominent in the northern portion.

On August 20, 1979, aquatic weeds were sparse to moderately dense in water depths to six feet. Robbins and curly pondweeds were most abundant with white water lily and spatterdock moderately dense at the northeastern and southern ends of the lake. The principal weed problems occur at the northeastern end near the inlets. The alga *Oscillatoria rubescens* was noticeable in water samples at the thermocline.

Mt. Tom Pond has been stocked with brown and rainbow trout.

Access is through a State boat launching area off Rte. 202. Fishing from shore is allowed in Mt. Tom State Park.

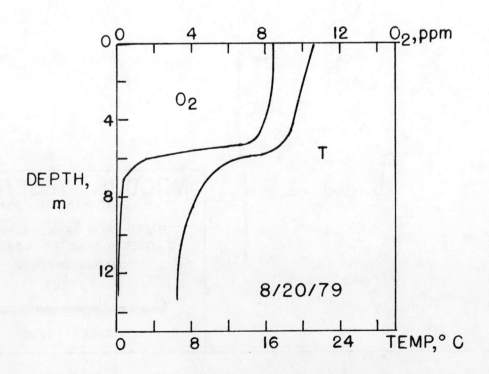

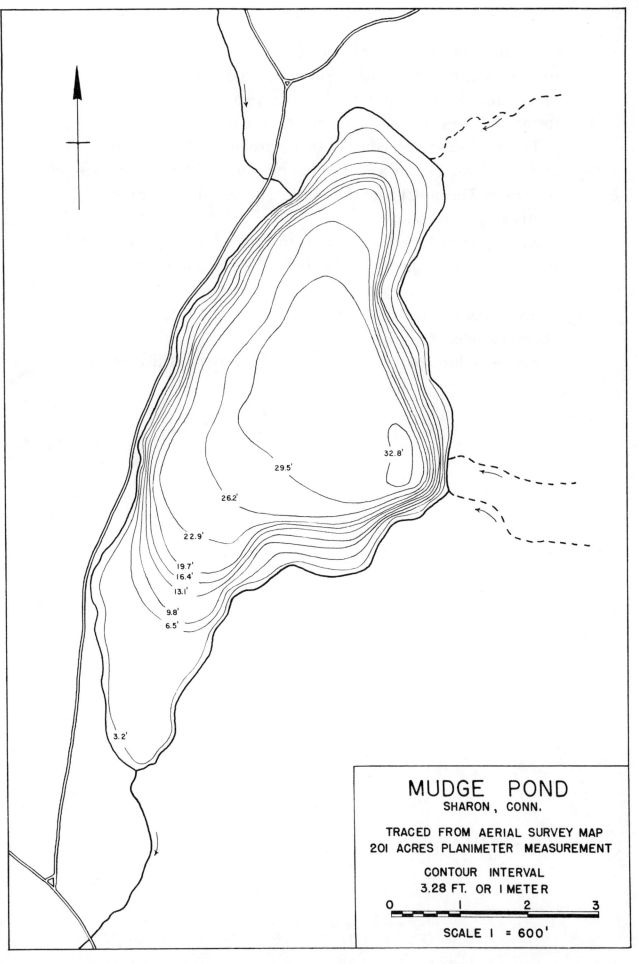

NORTH FARMS RESERVOIR

North Farms Reservoir is located in New Haven County in the town of Wallingford. It has a surface area of 62.5 acres, a maximum depth of 5 feet and an average depth of 3.1 feet. It was formed by construction of a dam across Catlin Brook.

The 473.6-acre watershed is 16.9% urban, 34.0% agricultural and 49.1% wooded or wet. The reservoir is fed by Catlin Brook to the north. The outlet is to the south. The landscape is primarily gently sloping terrain.

North Farms Reservoir appears to have an abundance of weeds most of the year. On March 26, 1979, the filamentous alga *Oedogonium* covered the bottom. On September 1, 1979, the entire lake contained a dense growth of coontail with watermeal abundant in coves.

Access is through a State boat launching area off Rte. 68.

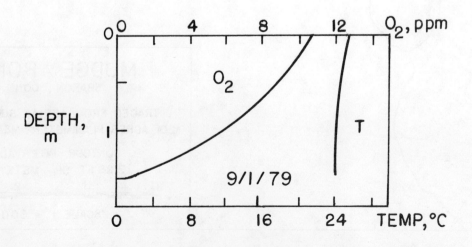

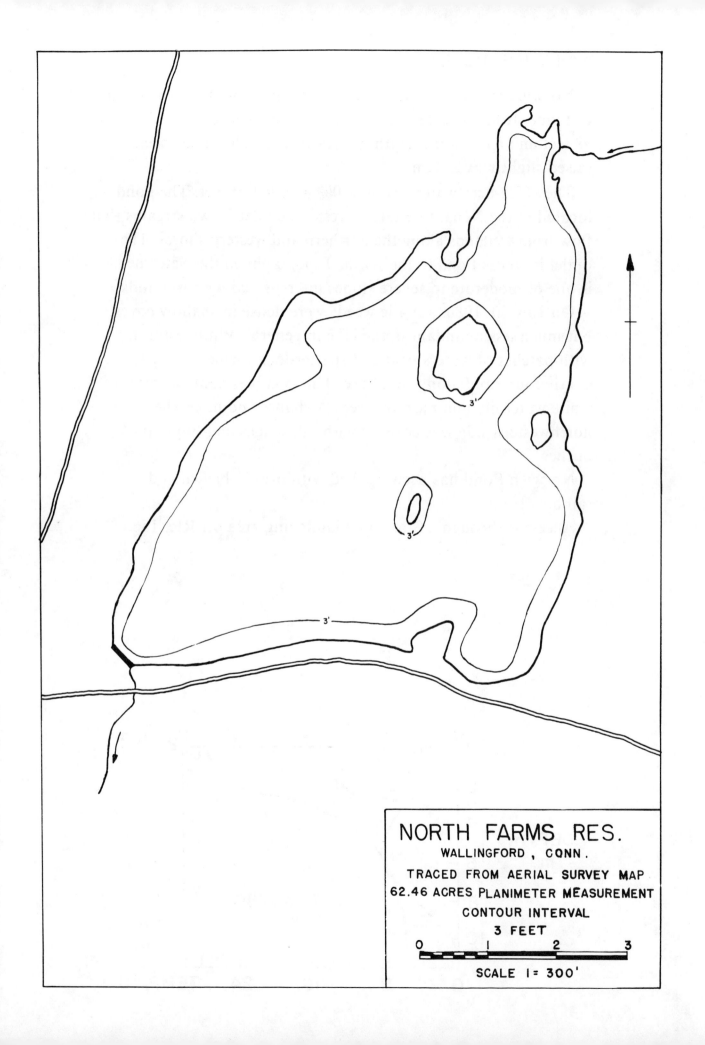

NORWICH POND

Norwich Pond is located in New London County in the town of Lyme. It has a surface area of 27.5 acres, a maximum depth of 38 feet and an average depth of 22.9 feet. The level has been raised slightly by a dam.

The 307.2-acre watershed is 100% wooded or wet. The pond is located in the Nehantic State Forest. The inlet is two streams that flow from wetlands along the northern and western shores. The outlet is Uncas Pond to the south. Topography in the watershed is one of moderate to severe slopes interspersed with wetlands.

On July 30, 1980, aquatic weeds were dense in shallow coves, but much less abundant in the lake in general. White water lily with patches of watershield and spatterdock formed a nearly continuous band along the shore. The inlet was congested with white water lily and pickerelweed. A shallow point on the northwestern side was covered with tall sedges and bog-type bushes.

Norwich Pond has been stocked with brook, brown and rainbow trout.

Access is through a State boat launching area off Rte. 156.

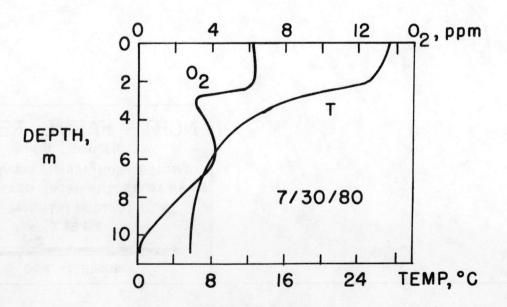

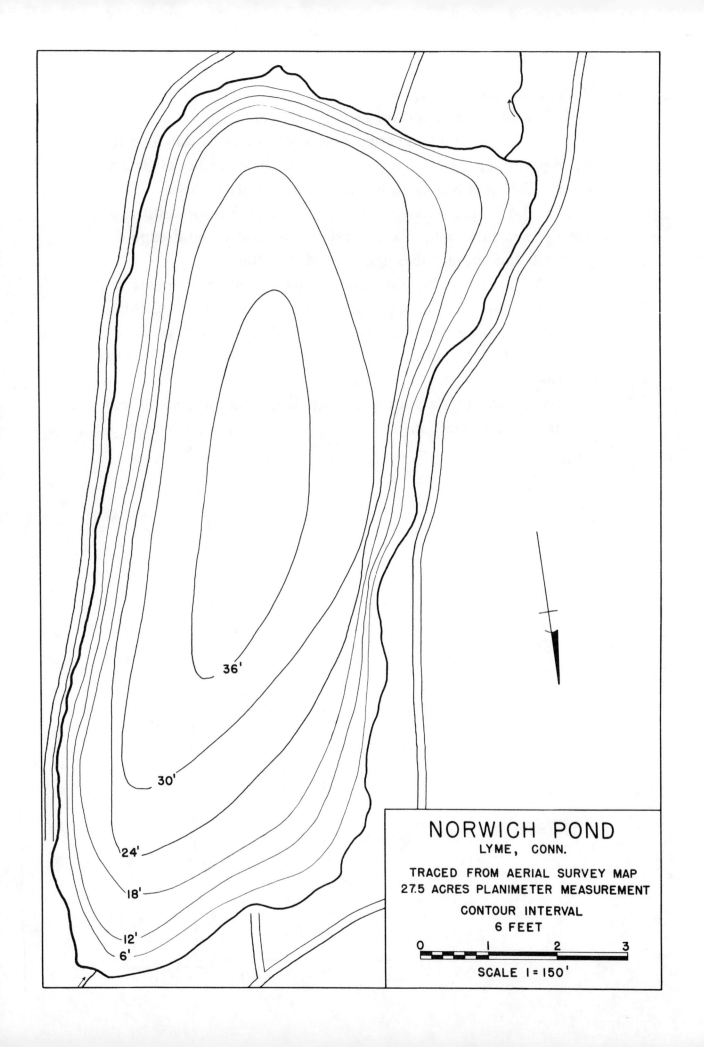

PACHAUG POND

Pachaug Pond is located in New London County in the town of Griswold. It has a surface area of 830.9 acres, a maximum depth of 18 feet and an average depth of 6.1 feet. It was formed by impounding the Pachaug River, which is the outlet of Beach Pond. The pond is located in the Pachaug State Forest.

The 33,408-acre watershed is 1.9% urban, 6.7% agricultural and 91.4% wooded or wet. The watershed has a varied landscape with generally moderate slopes and extensive wetlands.

On August 14, 1979, aquatic weeds were moderately dense in shallow areas. Some water milfoil was found near the launching area at the southern end and scattered beds of pickerelweed, bushy pondweed, spatterdock and pipewort were present in areas less than 3 feet deep.

Access is through a State boat launching area on the southern side in Pachaug State Forest.

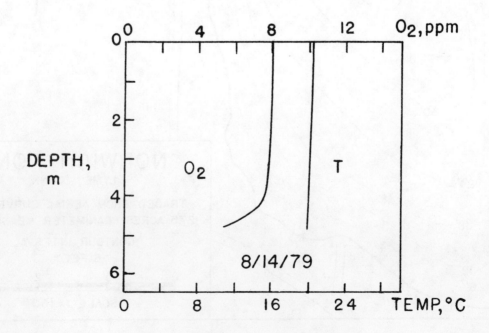

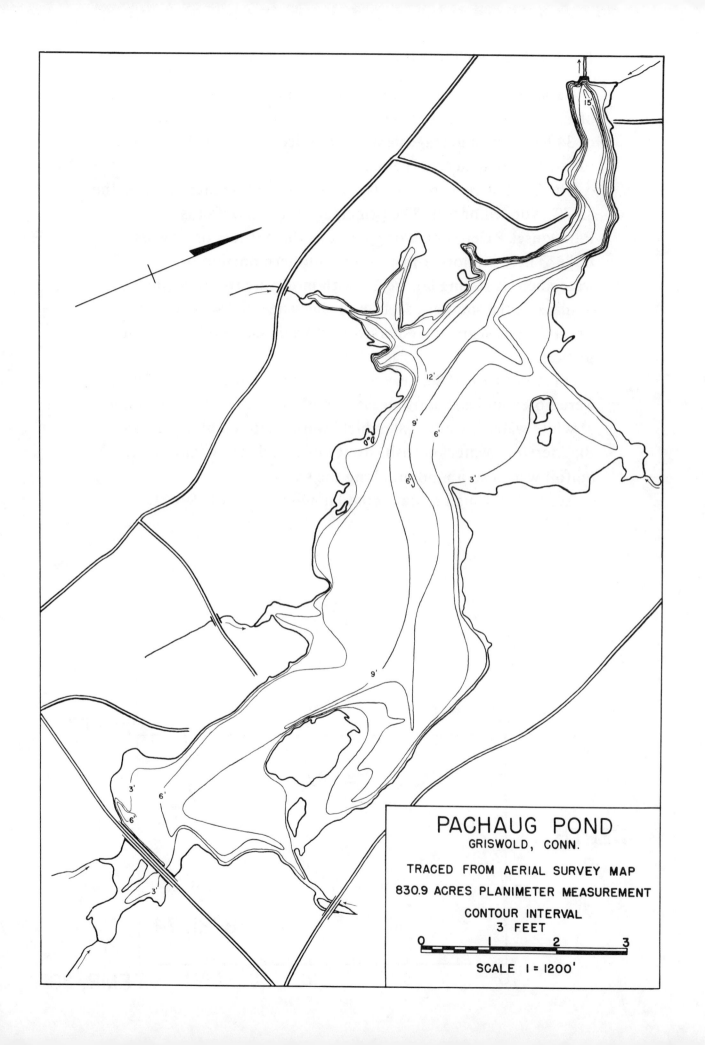

PATAGANSET LAKE

Pataganset Lake is located in New London County in the town of East Lyme. It has a surface area of 123 acres, a maximum depth of 34 feet and an average depth of 12.4 feet. The level has been raised by a dam across the outlet.

The 2,502-acre watershed is 11.7% urban, 4.6% agricultural and 83.7% wooded or wet. The principal inlet and outlet is the Pataganset River, which originates as the outflow of Powers Lake. There is another inlet in the southern portion of the watershed. A rolling landscape with moderate to severe slopes dominates the southern section of the watershed with a more irregular landscape interspersed with wetlands in the northern section.

On August 14, 1980, very dense growths of white water lily were noted in the shallow areas at both inlets. Other dense stands of white water lily and watershield were scattered along the shore. Bladderwort, water bulrush, stonewort, wild celery, and water milfoil were also abundant along the shore.

Access is through a State boat launching area off Rte. 1.

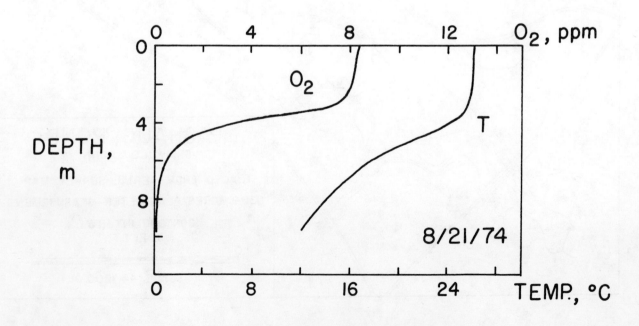

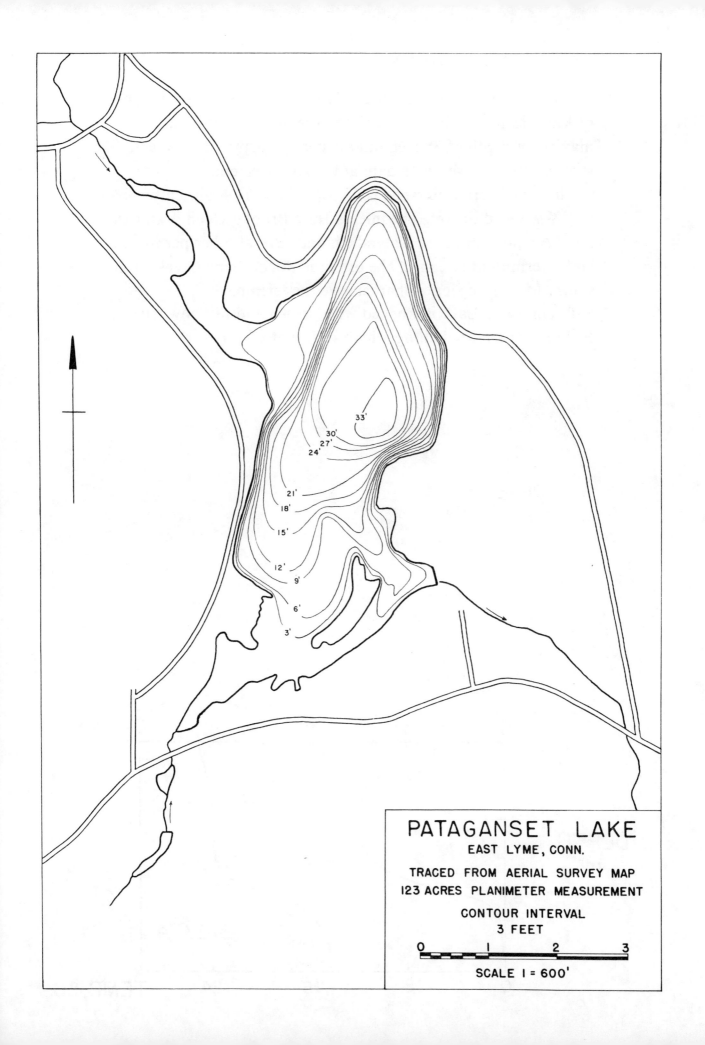

POCOTOPAUG LAKE

Pocotopaug Lake is located in Middlesex County in the town of East Hampton. It has a surface area of 511.7 acres, a maximum depth of 38 feet and an average depth of 11.3 feet. The level has been raised by a dam at the southern end.

The 2,893-acre watershed is 18.5% urban, 0.8% agricultural and 80.7% wooded or wet. The inlet is Hale Brook, which drains the northern portion of the watershed, and several other permanent and intermittent streams. The outlet is Pocotopaug Creek to the south. Moderate slopes characterize the terrain.

Pocotopaug Lake is stocked with brown and rainbow trout. Fishing is permitted along a highway right-of-way.

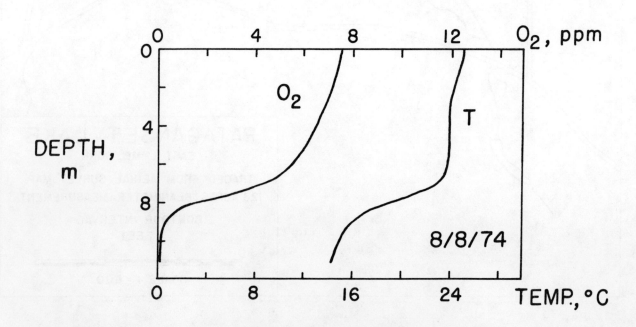

POWERS LAKE

Powers Lake is located in New London County in the town of East Lyme. It has a surface area of 152.6 acres, a maximum depth of 13 feet and an average depth of 7 feet. It has been impounded by a dam on the southern shore.

The 614.4-acre watershed is 100% wooded or wet. The inlet is several unnamed streams that drain the wetlands. An undulating landscape with moderate to severe slopes characterizes the watershed.

On July 21, 1980, aquatic weeds were moderately abundant. Water milfoil was the predominant species. Pipewort was common where the water depth did not exceed three feet and white water lily and bur reed inhabited the inlets. Other emergent species found were pickerelweed, arrowhead and spatterdock.

Access is through a State boat launching area off Rte. 1 near Pataganset Lake.

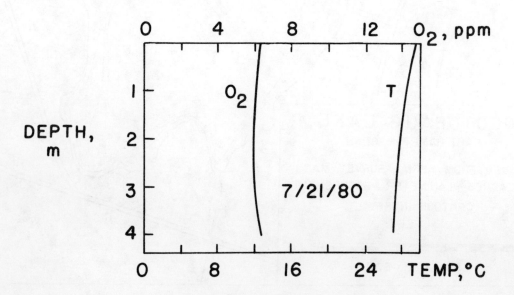

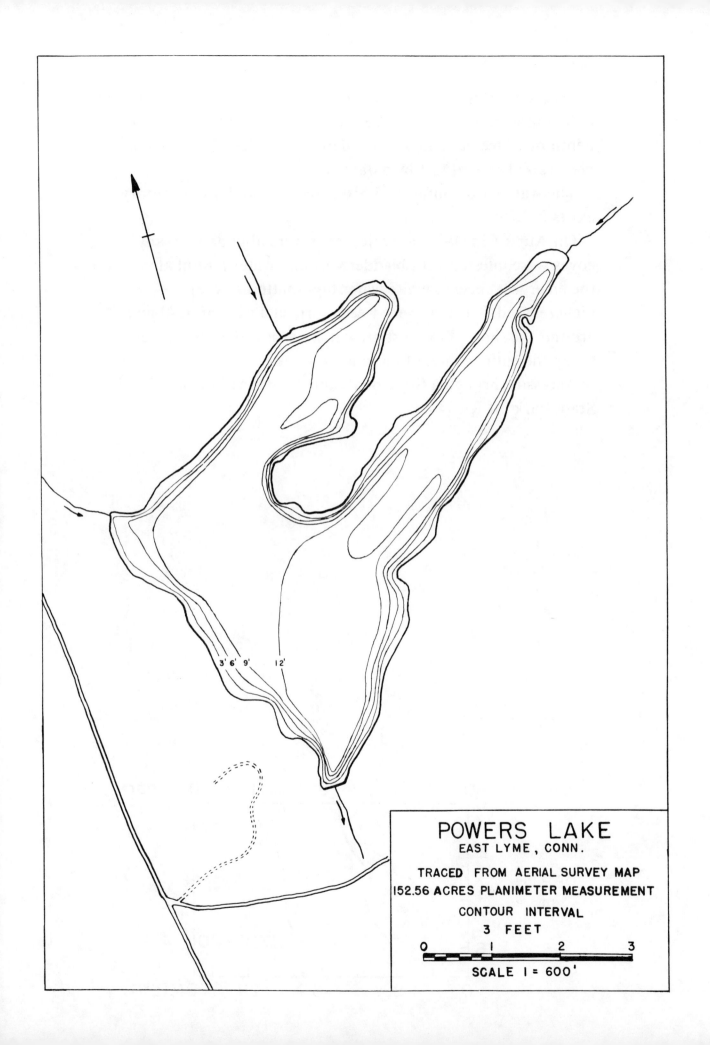

QUADDICK RESERVOIR

Quaddick Reservoir is located in Windham County in the town of Thompson. It has a surface area of 466.8 acres, a maximum depth of 25 feet and an average depth of 6.4 feet. The level has been raised several feet by a dam.

The watershed contains 15,616 acres. No land use inventory exists.

On August 25, 1980, aquatic weeds were abundant. Many coves were infested with bladderwort and water milfoil along with the emergent species, white water lily, spatterdock and pickerelweed. The emergent weeds formed a continuous band around the shore. Pipewort grew in areas less than three feet deep along the entire shore, often in large stands.

Access is through a State boat launching area in Quaddick State Park off Rte. 44.

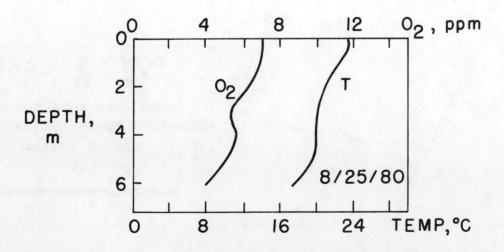

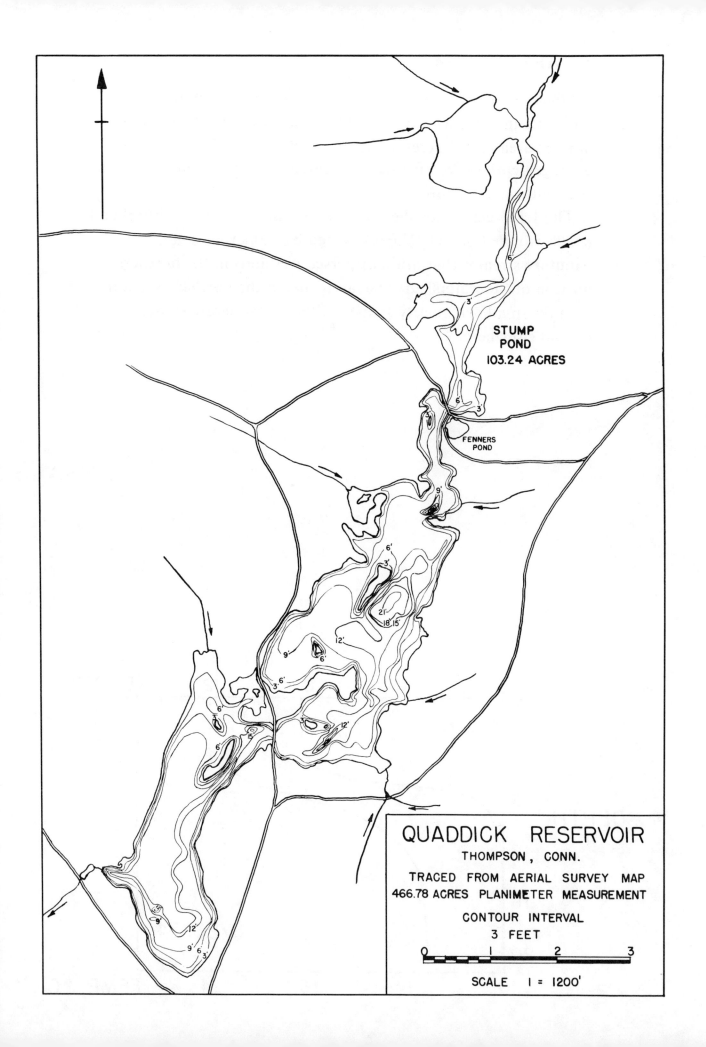

QUASSAPAUG LAKE

Quassapaug Lake is located in Litchfield and New Haven Counties in the towns of Woodbury and Middlebury. It has a surface area of 271 acres, a maximum depth of 65 feet and an average depth of 28.5 feet. The level has been raised by a dam on the southeastern shore.

The 1,165-acre watershed is 13.0% urban, 2.2% agricultural and 84.8% wooded or wet. The lake is fed by numerous small tributary streams that drain dispersed wetlands in the northern portion of the watershed. The landscape in the northern section features small, irregularly shaped hillocks. Moderate slopes prevail elsewhere.

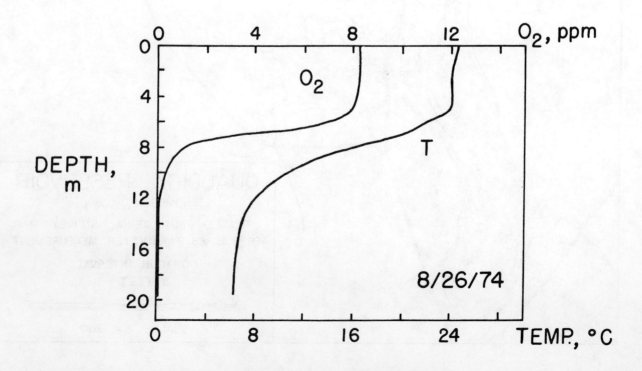

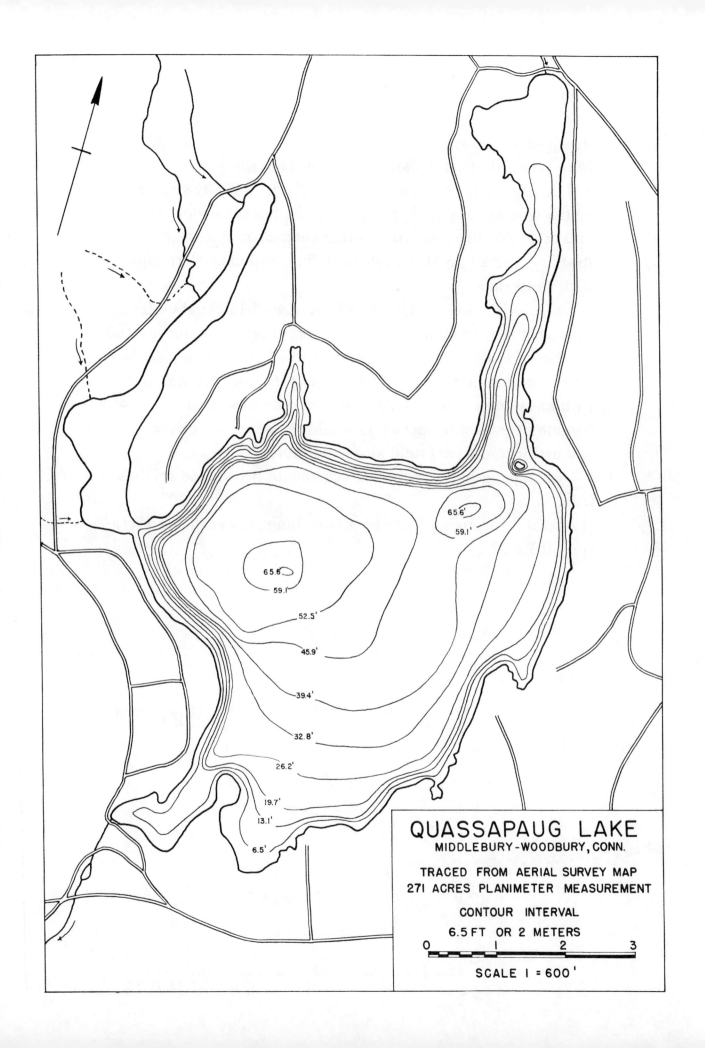

QUONNIPAUG LAKE

Quonnipaug Lake is located in New Haven County in the town of Guilford. It has a surface area of 111.6 acres, a maximum depth of 48 feet and an average depth of 13.6 feet. The level has been raised by a small dam on the southern shore.

The 1,683-acre watershed is 6.1% urban, 10.9% agricultural and 83.0% wooded or wet. The lake is fed by Sucker Brook from the east and two unnamed streams that enter from the north after draining a small pond. The surrounding landscape is irregular, with steep slopes.

On July 16, 1980, aquatic weeds were moderately abundant. Emergent weeds included extensive areas of white water lily and patches of spatterdock in the long, shallow cove leading to the outlet and also at the northern end near the inlet. Bigleaf and Robbins pondweeds were dense in water depths up to 9 feet, with common elodea interspersed in shallower water. Pickerelweed and arrowhead were found along the shore.

Quonnipaug Lake has been stocked with brown and rainbow trout.

Access is through a State boat launching area at the northern end off Rte. 77.

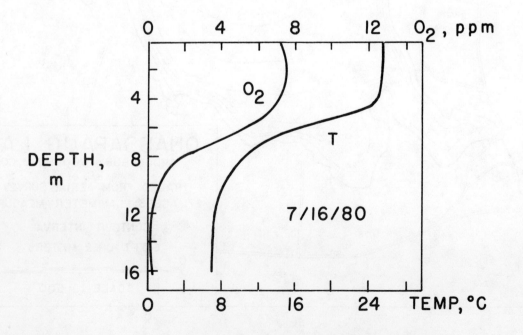

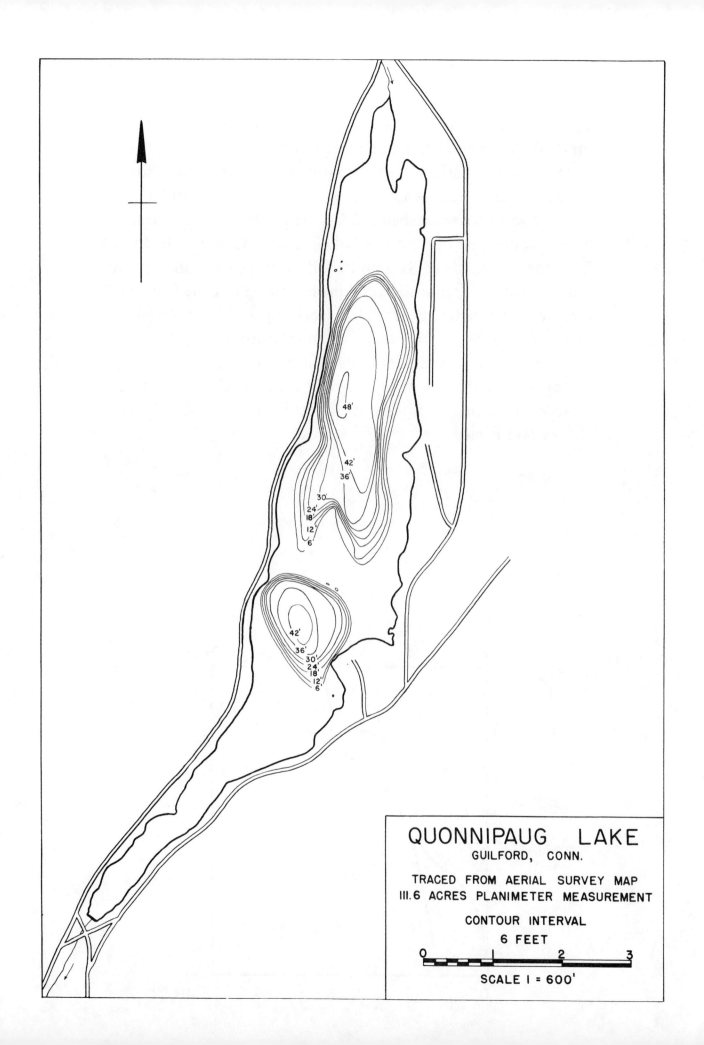

ROGERS LAKE

Rogers Lake is located in New London County in the towns of Lyme and Old Lyme. It has a surface area of 264.9 acres, a maximum depth of 66 feet and an average depth of 20.1 feet. The lake consists of two distinct basins separated by a band of shallow water. The level has been raised by a dam across the outlet.

The 4,819-acre watershed is 5.9% urban, 1.3% agricultural and 92.8% wooded or wet. Rogers Lake is fed by Grassy Hill Brook from the north, Broad Swamp Brook from the east, and several other unnamed streams. Several prominent landscape features include Garre Ledge, Becket Hill, Hart Hill and Cranberry Ledge. Generally, severe slopes are found throughout the watershed with some moderation of slope near the lake.

Rogers Lake has been stocked with brown and rainbow trout.

Access is through a State boat launching area off Rte. 1 on Grassy Hill Road.

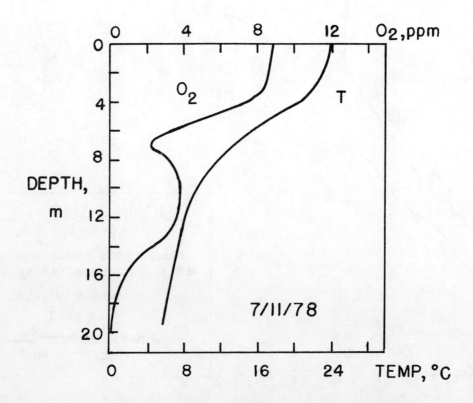

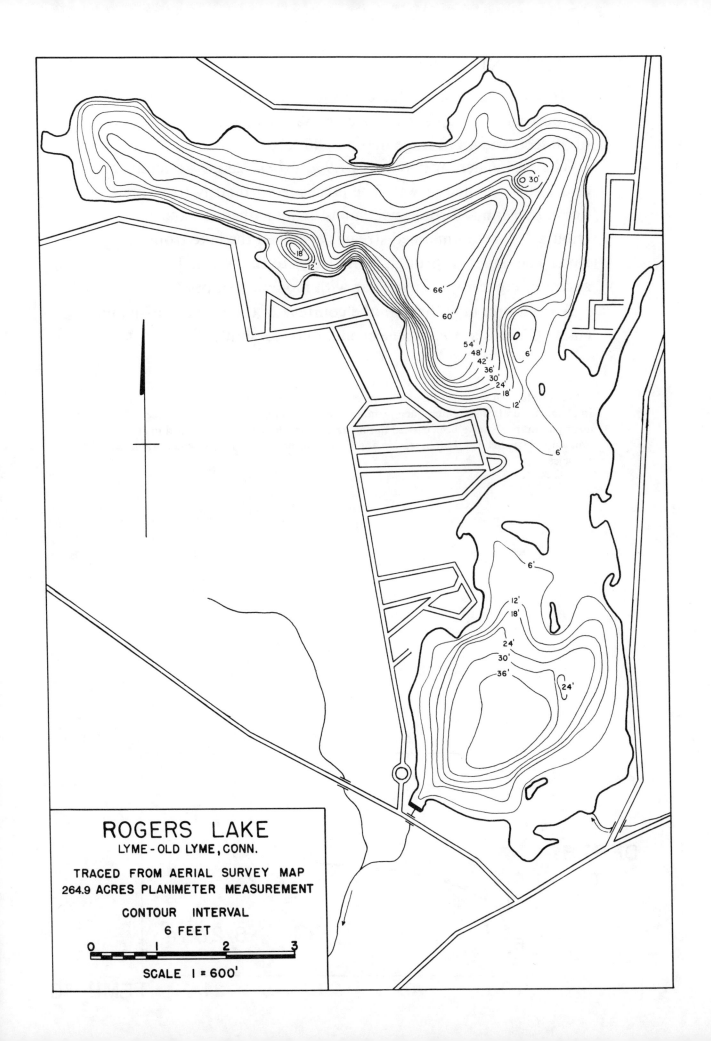

ROSELAND LAKE

Roseland Lake is located in Windham County in the town of Woodstock. It has a surface area of 88 acres, a maximum depth of 20 feet, and an average depth of 10 feet.

The 19,456-acre watershed is 1.9%, urban, 25.2% agricultural and 72.9% wooded or wet. The lake is fed primarily from the north by Muddy Brook, which has many named streams as tributaries. Two unnamed streams also feed the lake from the west. The outlet is through Little River to the south. The surrounding terrain is irregular with moderate slopes.*

Access, provided through the courtesy of the Putnam Fish and Game Club, is at the southern end, outside South Woodstock.

*This watershed was the subject of a special study under a USDA/EPA 208 Model Implementation Project. Subsequently, many practices to reduce non-point sources of sediment and nutrients have been carried out on agricultural land in the watershed in cooperation with the Agricultural Stabilization and Conservation Service and other state and federal agencies.

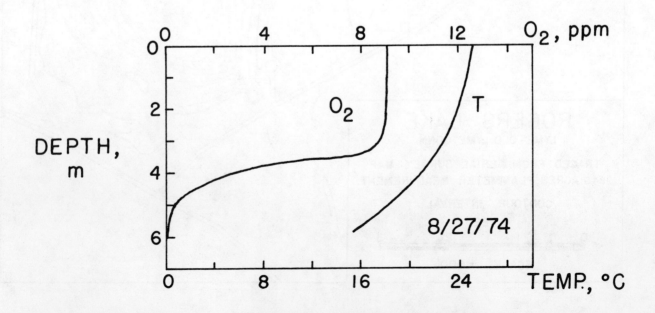

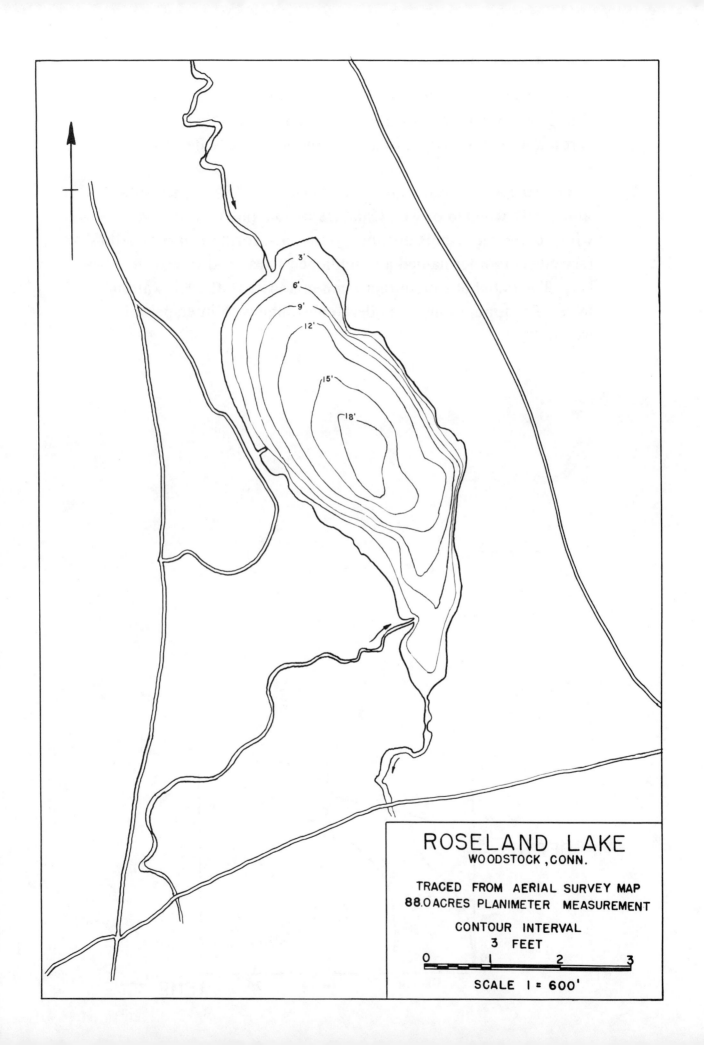

SHENIPSIT LAKE

Shenipsit Lake is located in Tolland County in the towns of Ellington, Tolland and Vernon. It has a surface area of 522.8 acres, a maximum depth of 68 feet and an average depth of 30 feet.

The 10,496-acre watershed is 10.4% urban, 7.9% agricultural and 81.7% wooded or wet. Charters Brook and West Brook, which converge a short distance to the east, form the largest tributary. Several other named and unnamed streams also feed the lake. The outlet at the southern end gives rise to the Hockanum River. The topography is moderately rolling with interspersed wetlands.

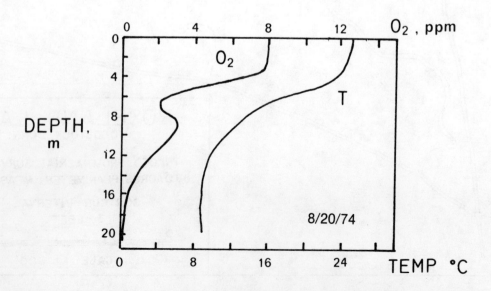

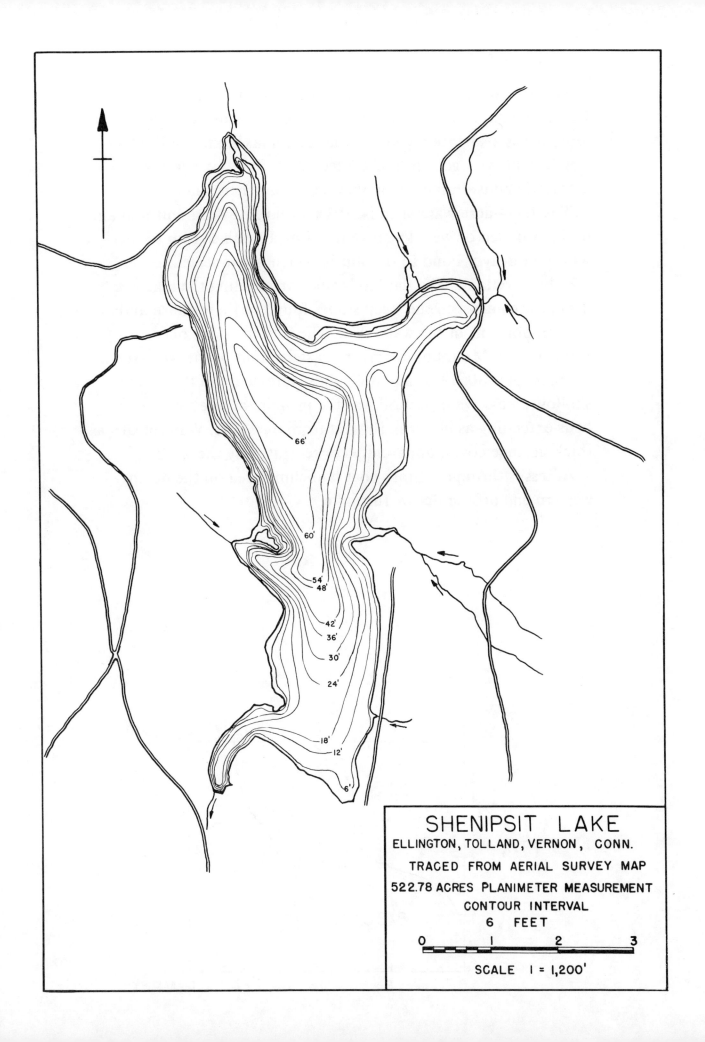

SILVER LAKE

Silver Lake, also known as Peat Works Pond, is located in Hartford and Middlesex Counties in the towns of Berlin and Meriden. It has a surface area of 151 acres, a maximum depth of 12 feet and an average depth of 4.5 feet. This State-owned lake is artificial and is drained by a dam on the northern shore.

The 1,274-acre watershed is 30.3% urban, 1.6% agricultural and 68.1% wooded or wet. The lake is fed by a small stream to the west and Beaver Pond to the south, which is connected to the lake by a wetland. The flat landscape surrounding the lake is a dramatic contrast to the steep slopes found in the western and eastern portions of the watershed. The western ridge of Lamentation Mountain forms the eastern limit of the watershed.

On August 30, 1979, aquatic weeds were dense in the extensive shallow areas. Water milfoil was the prominent weed with dense beds extending as much as 100 feet from the shore. Watermeal was thick in some coves, apparently concentrated by the wind.

Access is through a State boat launching area on the northwestern end off the Berlin Turnpike.

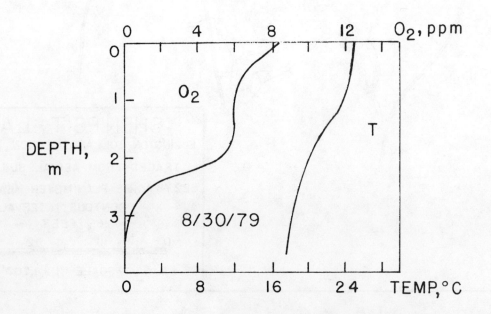

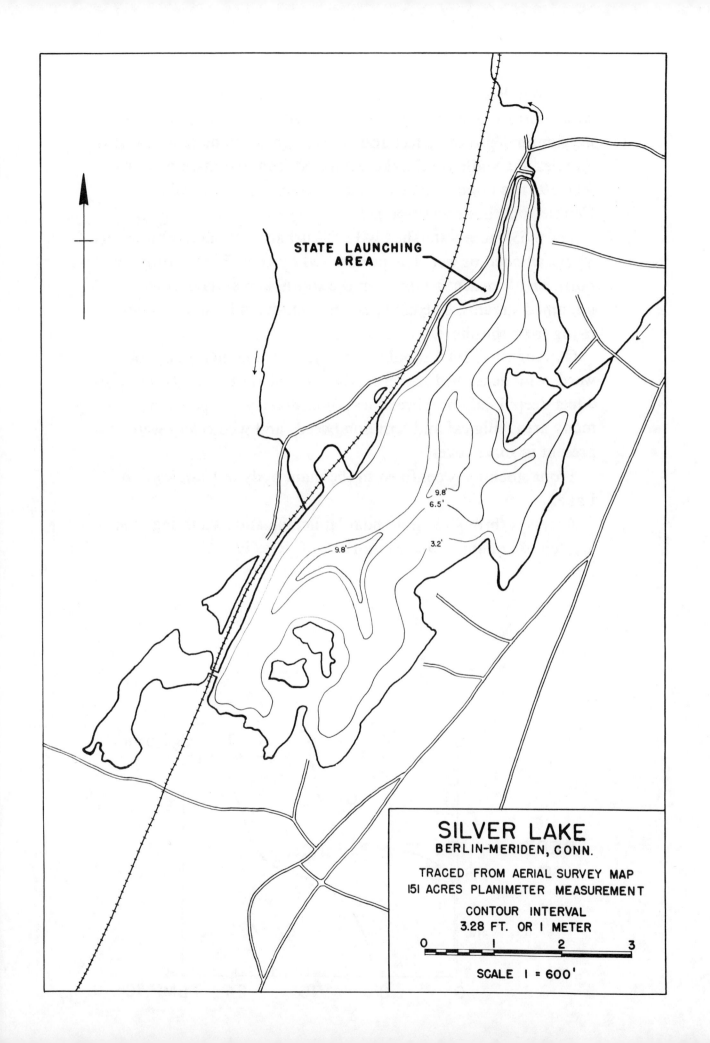

SQUANTZ POND

Squantz Pond is located in Fairfield County in the towns of New Fairfield and Sherman. It has a surface area of 288 acres, a maximum depth of 47 feet and an average depth of 22.9 feet. It is an inlet of Candlewood Lake separated from the main body by Rte. 39. The pond is natural, but the level was raised when Candlewood Lake was flooded.

The 3,635-acre watershed is 15.5% urban, 2.4% agricultural and 82.1% wooded or wet. The pond is fed by Glen Brook from the north and Worden Brook from the south, and several other unnamed streams. The landscape is dominated by steep slopes rising from the shore.

On August 7, 1980 coontail was the most abundant aquatic weed, with dense beds covering most of the bottom in water up to 9 feet deep. Leafy pondweed was intermixed with the coontail in many areas. Bigleaf and fern pondweeds and wild celery were present in some coves.

Trout stocking is confined to the main body of Candlewood Lake.

Access is through a State boat launching and swimming area located in Squantz Pond State Park off Rte. 39.

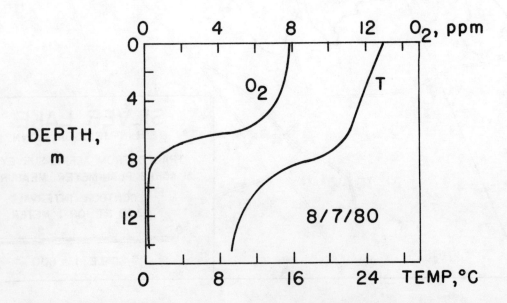

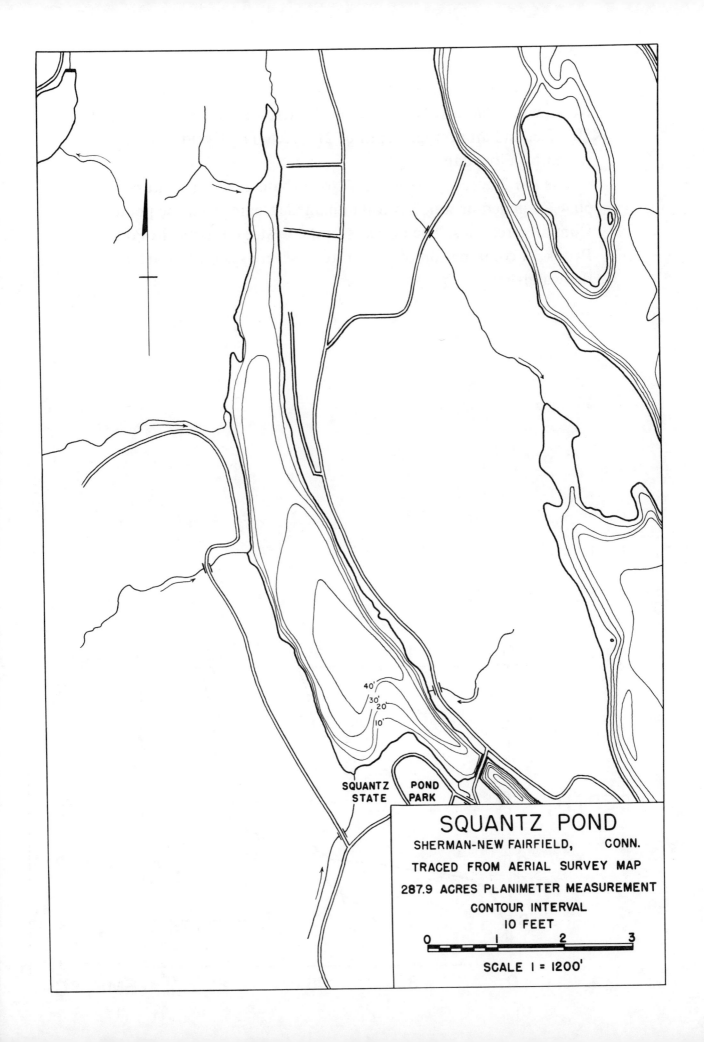

TAUNTON POND

Taunton Pond is located in Fairfield County in the town of Newtown. It has a surface area of 126 acres, a maximum depth of 29.5 feet and an average depth of 21.5 feet. The level has been raised by a low dam.

The 851.2-acre watershed is 47.6% urban, 2.0% agricultural and 50.4% wooded or wet. Several unnamed streams feed the pond from all directions. The outlet is to the west, giving rise to Pond Brook. A combination of moderate to severe slopes characterizes the watershed.

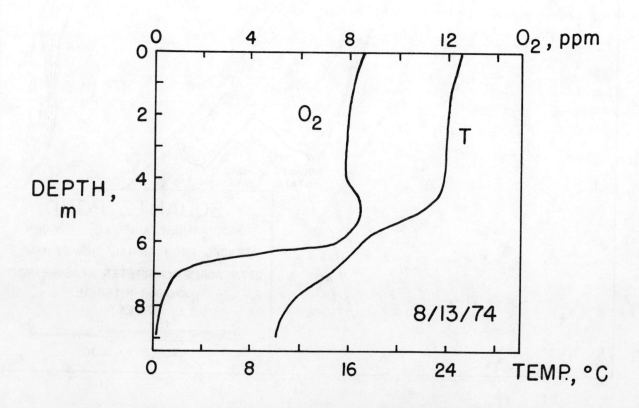

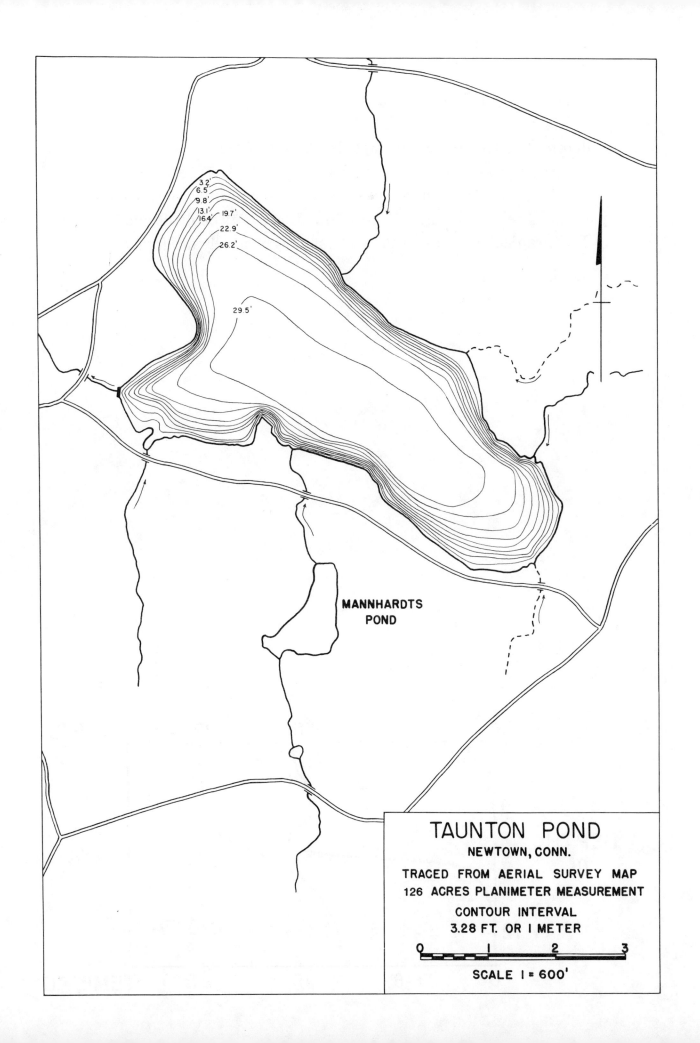

TERRAMUGGUS LAKE

Terramuggus Lake is located in Hartford County in the town of Marlborough. It has a surface area of 83 acres, a maximum depth of 43 feet and an average depth of 21.4 feet.

The 345.6-acre watershed is 63% urban and 37% wooded or wet. The outlet is an unnamed stream on the northwestern shore. Feeder streams and associated wetlands are noticeably absent. Topographic variation in the watershed is generally moderate.

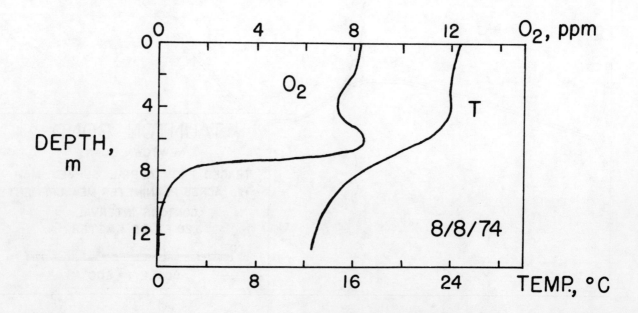

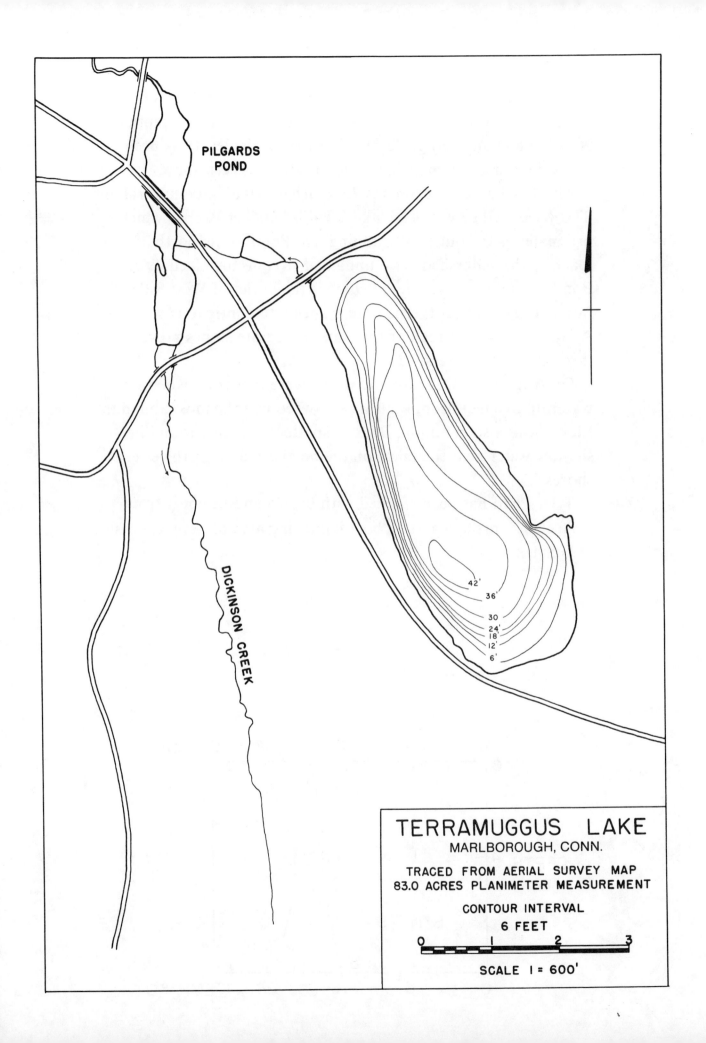

TYLER POND

Tyler Pond is located in Litchfield County in the town of Goshen. It has a surface area of 182 acres, a maximum depth of 26.2 feet and an average depth of 12.1 feet. The level has been raised by a small dam at the outlet on the southern shore.

The 4,045-acre watershed is 3.0% urban, 10.0% agricultural and 87.0% wooded or wet. The lake is fed by Sucker Brook, which originates as the outlet of Cunningham Pond in the northern reach of the watershed. Two unnamed streams also feed Tyler Pond from the north, one of which is the outlet of West Side Pond. The entire northern shore grades into contiguous wetlands. A landscape with a regular form and moderate slopes prevails throughout the watershed.

On August 17, 1979, aquatic weeds were moderately dense in water up to 6 feet deep. Bigleaf pondweed was the most abundant. Elodea was also common. Beds of spatterdock were found in shallow water near the inlets on the northern and northwestern shores.

Tyler Pond has been stocked with brown and rainbow trout.

Access is through a State boat launching area off Rte. 4, outside of Goshen.

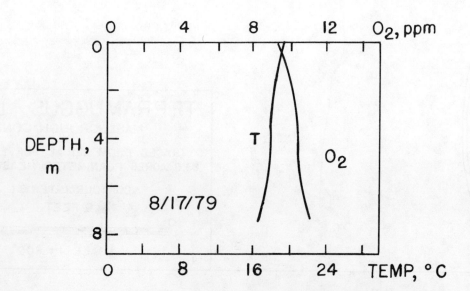

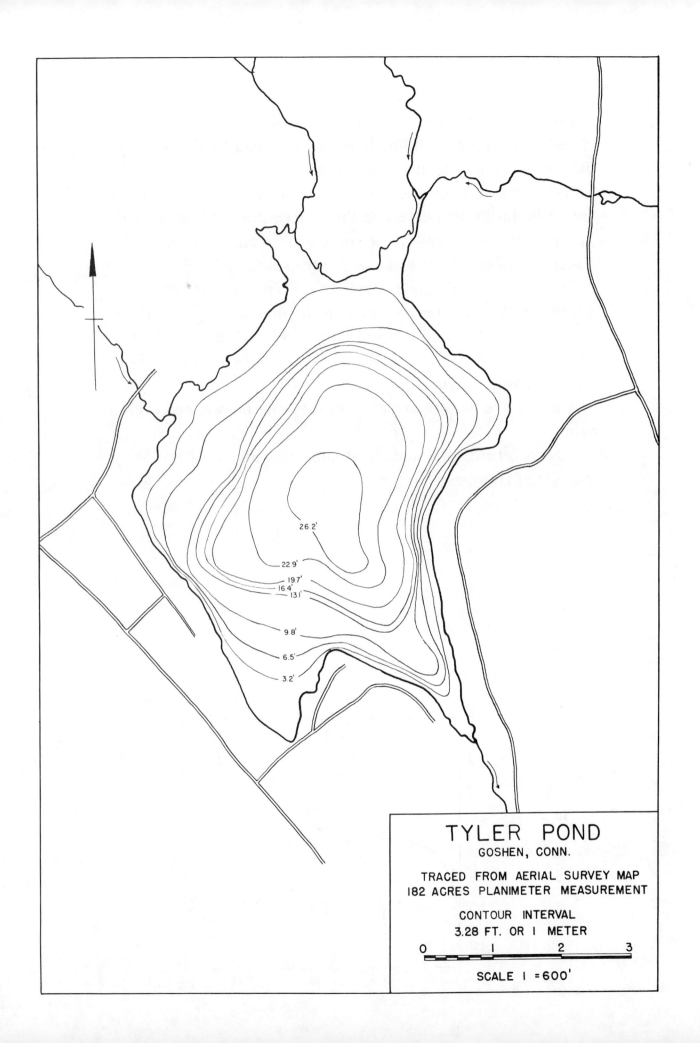

UNCAS LAKE

Uncas Lake, also known as Hog Pond, is located in New London County in the town of Lyme. It has a surface area of 69 acres, a maximum depth of 40 feet and an average depth of 22.2 feet. It is in the Nehantic State Forest.

The 960-acre watershed is 100% wooded or wet. The lake is fed by a tributary which enters on the eastern shore. Moderate to severe slopes are found in the entire watershed, with the exception of small wetland areas to the south and west.

On July 30, 1980, aquatic weeds were abundant at only a few locations. White water lily, watershield and various pondweeds were found occasionally in water up to 6 feet deep. Pipewort was common in shallower water and pickerelweed grew in small patches along the shore.

Uncas Lake has been stocked with brook, brown, and rainbow trout.

Access is through a State boat launching area off Rte. 156, outside of Hamburg.

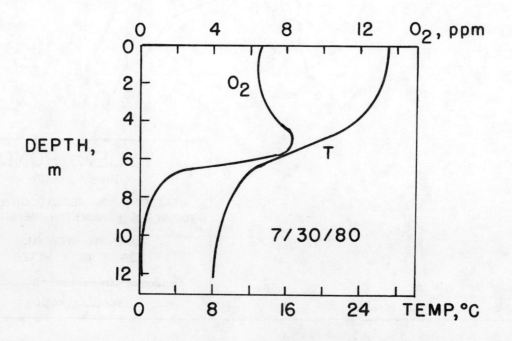

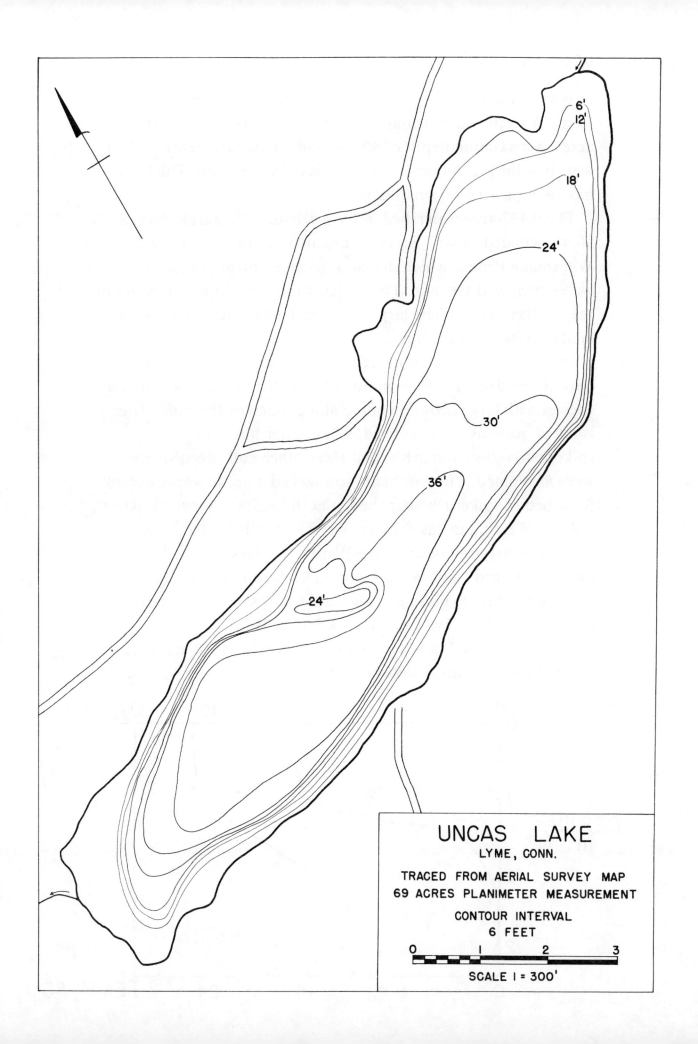

WARAMAUG LAKE

Waramaug Lake is located in Litchfield County in the towns of Kent, Warren and Washington. It has a surface area of 680.2 acres, a maximum depth of 40 feet and an average depth of 22.1 feet. It is the second largest natural lake in the State. The level has been raised by a small dam.

The 9,152-acre watershed is 3.9% urban, 5.7% agricultural and 90.4% wooded or wet. Several unnamed streams and Lake Waramaug Brook, which drains a sizeable portion of the watershed, feed the lake. The outlet is the East Aspetuck River on the southern shore. The large watershed is characterized by a hilly landscape with steep slopes.

In August 1980, aquatic weed growth was limited to areas less than 10 feet deep. Robbins pondweed was the single most abundant species with large beds extending along much of the shore. Small isolated growths of arrowhead, white water lily, coontail, spikerush, yellow water lily, and three other pondweed species were also noted. The southeast arm lacked aquatic weeds except for a bed of spikerush near the outlet to the East Aspetuck River.

Lake Waramaug has been extensively studied in recent years under the direction of the Lake Waramaug Task Force. The lake has experienced heavy algal blooms (Kortmann, et al., 1982) and an experimental hypolimnetic withdrawal system is currently being tested (Lake Waramaug Task Force, 1978).

Access is through a State boat launching area in Waramaug State Park at the northwestern end.

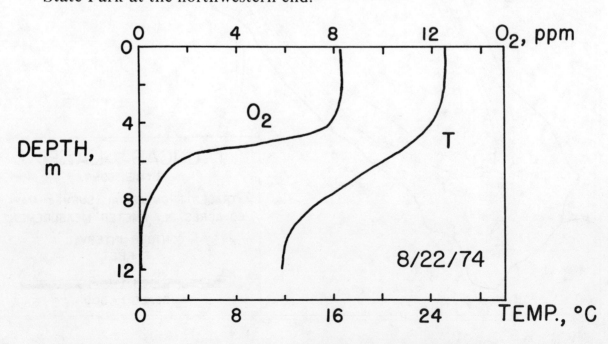

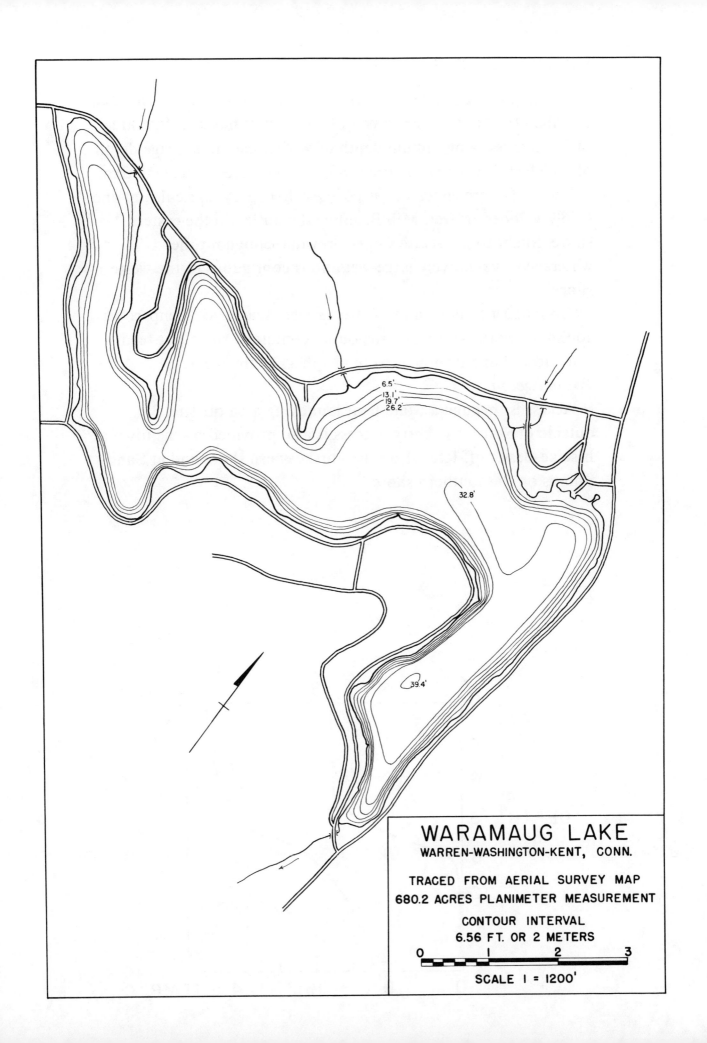

WAUMGUMBAUG LAKE

Waumgumbaug Lake, also known as Coventry Lake, is located in Tolland County in the town of Coventry. It has a surface area of 377.5 acres, a maximum depth of 40 feet and an average depth of 28.9 feet. The level has been raised several feet by a small dam.

The 2,125-acre watershed is 26.8% urban, 3.6% agricultural and 69.6% wooded or wet. Mill Brook is the outlet on the eastern shore. Slight to moderate slopes prevail throughout the watershed. A relatively large wetland is contiguous to the western shore.

On August 5, 1980, aquatic weeds were scarce. Goldenpert was found along the shore and the outlet contained some bur reed. The most abundant weeds were members of the genus *Potomogeton*.

Access is through a State boat launching area off Rte. 31 outside of South Coventry. Access is also provided by the town at Liscike Beach off Rte. 31 on the northeastern shore and at Sandy Shores on the southern shore.

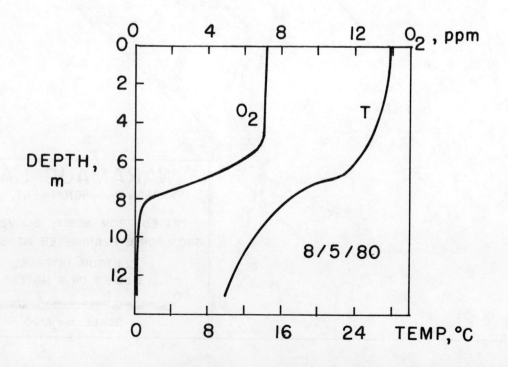

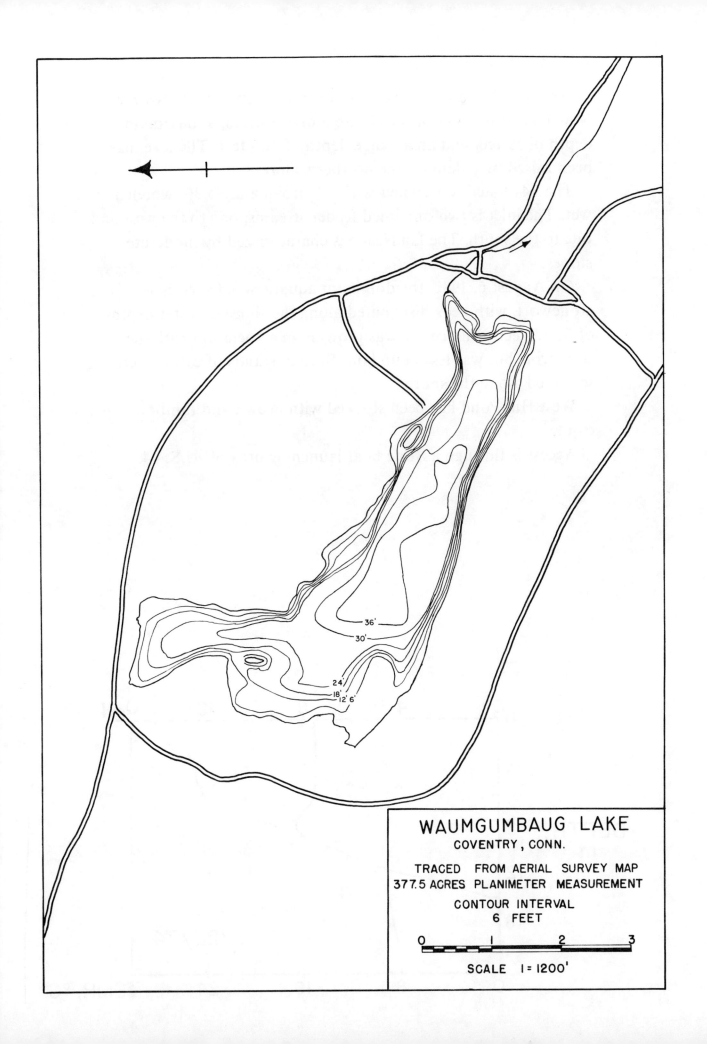

WEST HILL POND

West Hill Pond is located in Litchfield County in the town of New Hartford. It has a surface area of 238 acres, a maximum depth of 59 feet and an average depth of 31.8 feet. The level has been raised by a dam on the northern shore.

The 844.8-acre watershed is 15.7% urban and 84.3% wooded or wet. The inlet is two unnamed feeder streams, one to the west and one to the south. The landscape is characterized by moderate slopes.

On August 6, 1982, the dominant aquatic weed species was stonewort, with beds distributed along the shore in water depths of 5 to 9 feet. Wild celery was growing in association with the stonewort but was less abundant. Several stands of cattail were scattered along the shore.

West Hill Pond has been stocked with brown and rainbow trout.

Access is through a State boat launching area off U.S. 44.

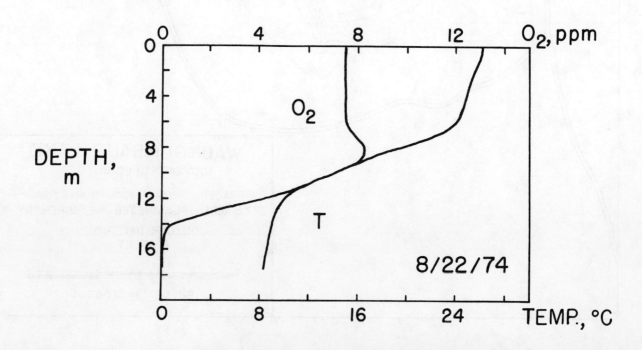

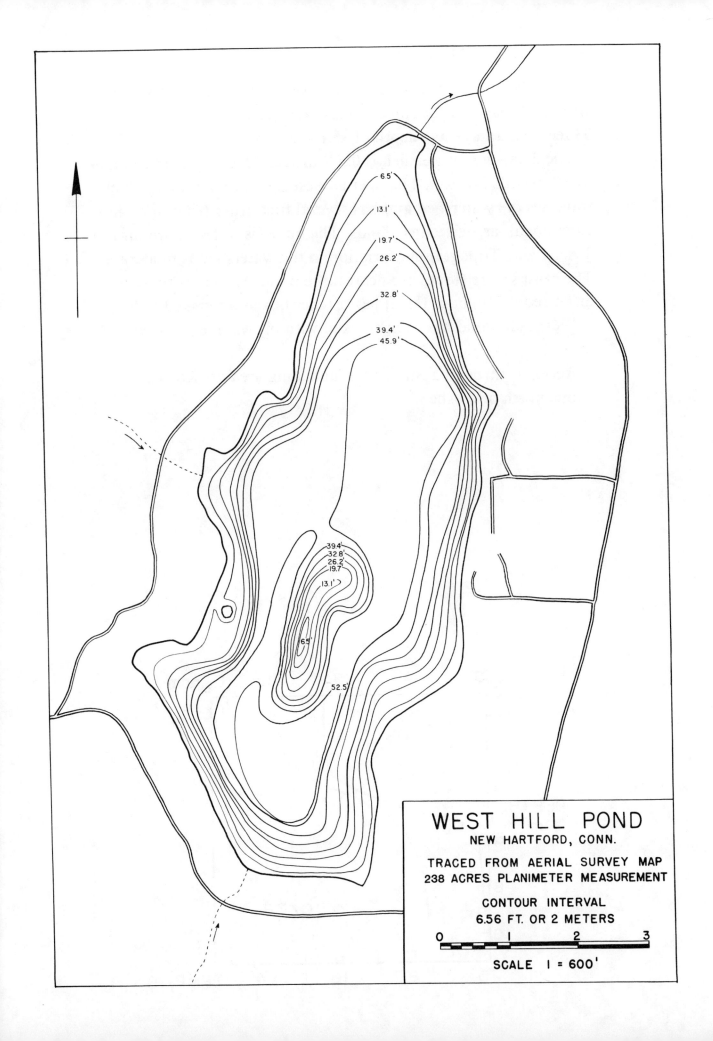

WEST SIDE POND

West Side Pond is located in Litchfield County in the town of Goshen. It has a surface area of 42.4 acres, a maximum depth of 33 feet and an average depth of 15.1 feet.

The 2,181-acre watershed is 0.3% urban, 12.4% agricultural and 87.3% wooded or wet. The inlet is West Side Pond Brook, which traverses the watershed and has several first order tributaries and streambank-associated wetlands. The outlet is to the south into Tyler Lake. Topographic variation in the watershed is moderate. The most severe slopes are nearest the pond. Wetlands are dispersed throughout the upper reaches of the watershed.

West Side Pond has been stocked with brown and rainbow trout.

Access is through a State boat launching area off Rte. 63, 1 mile north of Goshen.

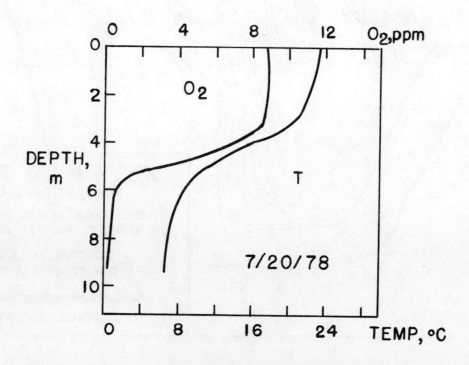

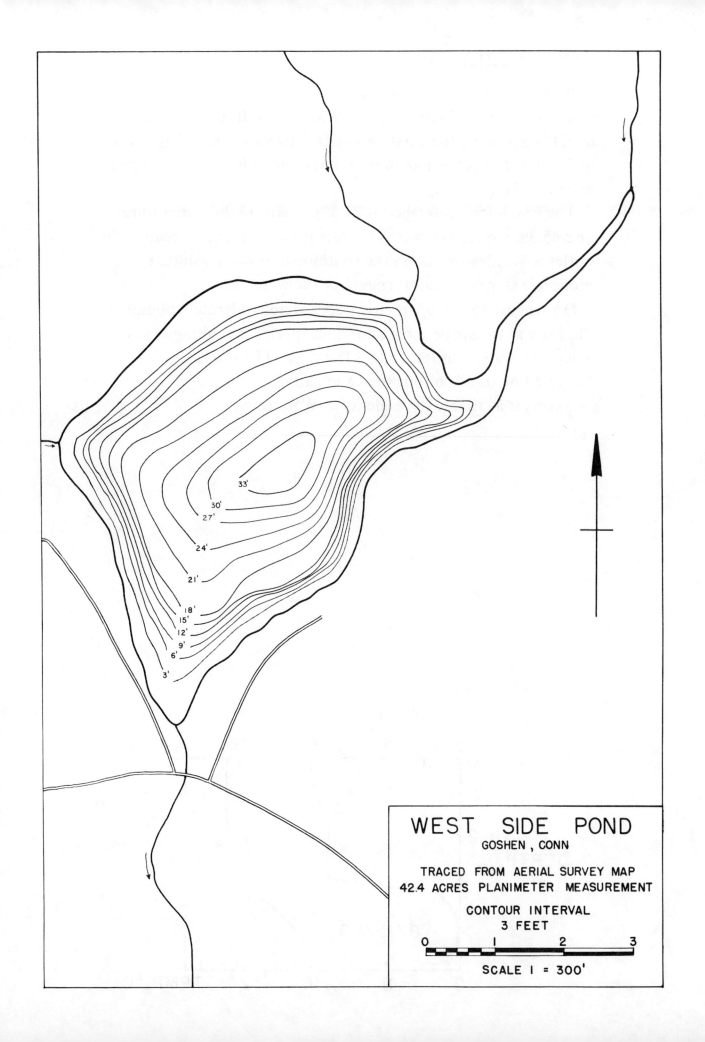

WINNEMAUG LAKE

Winnemaug Lake, also known as Wattles Pond, is located in Litchfield County in the town of Watertown. It has a surface area of 120 acres, a maximum depth of 16.5 feet and an average depth of 7.9 feet. It has been artificially impounded by an earthen and masonry dam.

The 684.8-acre watershed is 21.5% urban, 13.2% agricultural and 65.3% wooded or wet. The inlet is two tributary streams. The outlet is Wattles Brook on the southwestern shore. Slight to moderate slopes prevail throughout the watershed.

On August 15, 1980, aquatic weeds were moderately abundant. Elodea was the single most abundant species inhabiting coves where the water was less than 3 feet deep. The shore was bordered by stands of bur reed and cattail.

Access is available to residents of Watertown.

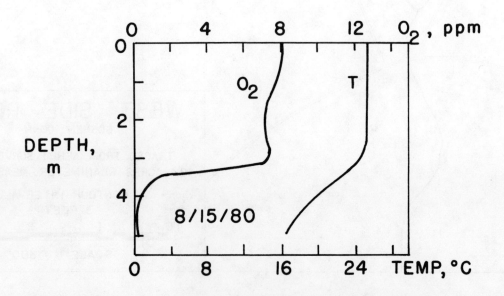

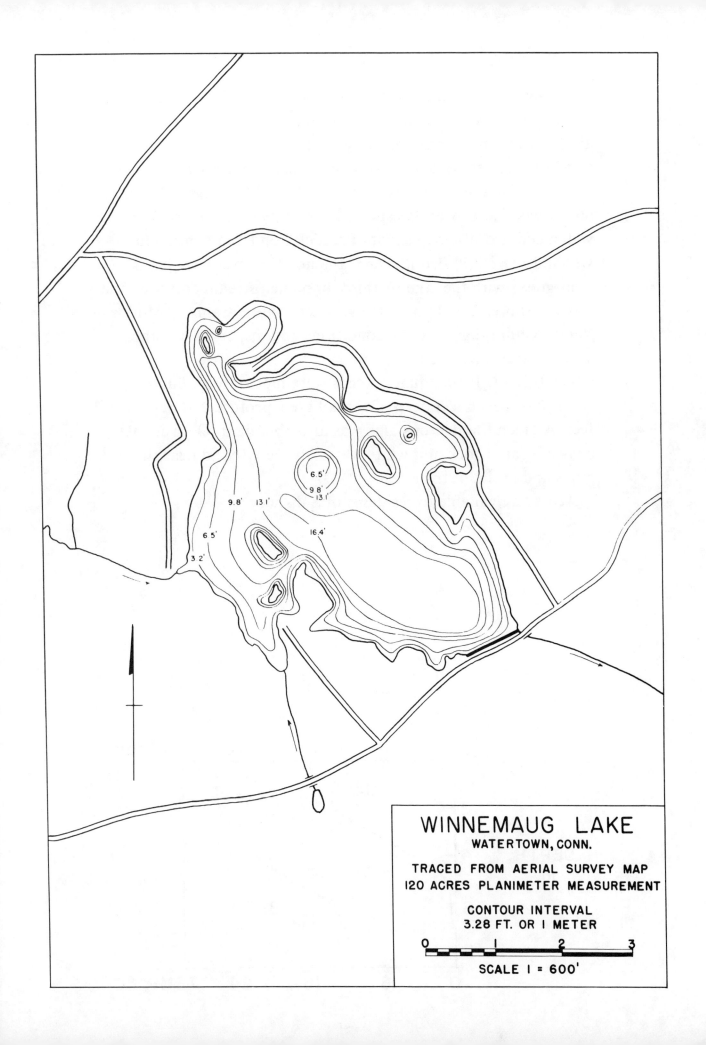

WONONPAKOOK LAKE

Wononpakook Lake, sometimes called Long Pond, is located in Litchfield County in the town of Salisbury. It has a surface area of 164 acres, a maximum depth of 24.5 feet and an average depth of 11.5 feet. The level has been raised by a dam.

The 2,778-acre watershed is 5.8% urban, 27.3% agricultural and 66.9% wooded or wet. Beeslick Brook drains the eastern watershed and originates from Beeslick Pond and contiguous wetlands. Ore Hill Brook drains the northern wetland that is contiguous with the lake. A third, unnamed stream enters on the western shore. The dam on the southern shore drains to Mudge Pond. A full range of slope conditions and dispersed wetlands exists in this watershed.

On July 18, 1980, aquatic weeds were moderately abundant in water less than 6 feet deep. Coontail grew profusely at depths to 3 feet. A green filamentous alga was also abundant. Occasional beds of spatterdock and white water lily were found near the inlets and outlet of the lake.

Access is available to residents of Salisbury.

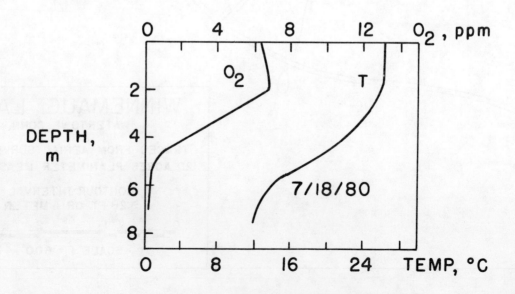

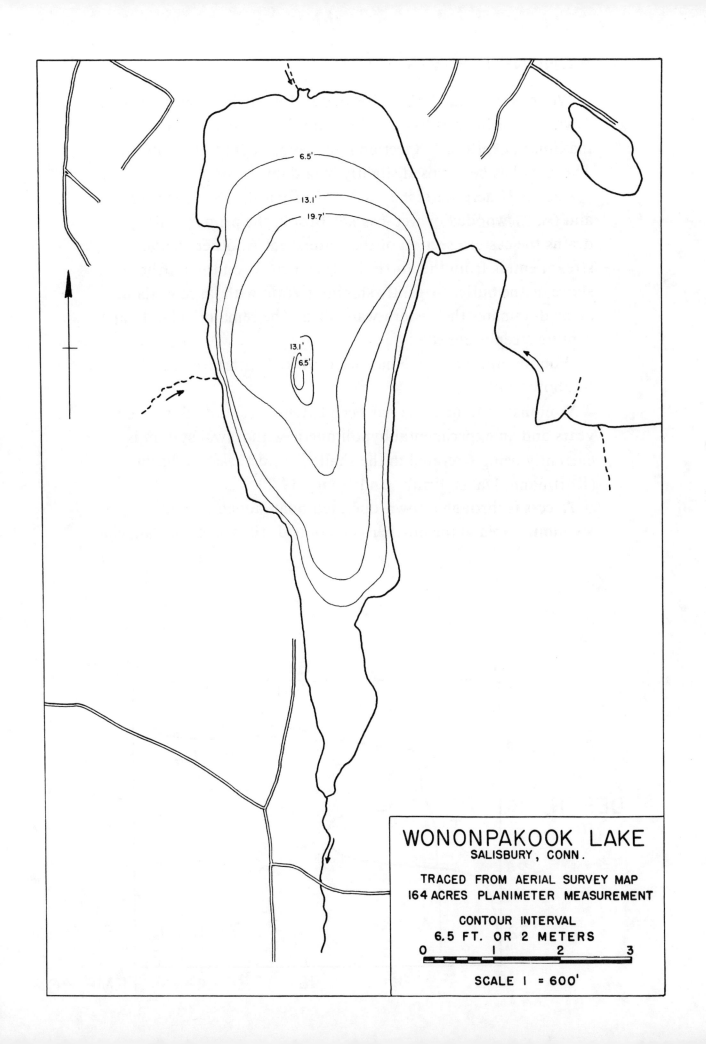

WONONSCOPOMUC LAKE

Wononscopomuc Lake is located in Litchfield County in the town of Salisbury. It has a surface area of 352.6 acres, a maximum depth of 108 feet and an average depth of 36.3 feet. The level has been raised slightly by a dam.

The 1,677-acre watershed is 21.0% urban, 10.3% agricultural and 68.7% wooded or wet. Sucker Brook, the primary inlet, drains the eastern section of the watershed. Another smaller stream enters from the north. Factory Brook, on the northern shore, is the outlet. A gently sloping terrain with large wetland areas dominates the eastern watershed. The remaining landscape is more steeply sloped.

Wononscopomuc Lake has been stocked with brown and rainbow trout.

Wononscopomuc Lake has been extensively studied in recent years and an experimental hypolimnetic withdrawal system is currently being operated in the shallower northeastern basin (Kortmann, Davis, Frink, and Henry, 1983).

Access is through a town-operated boat launching and swimming area at the northern end outside the village of Lakeville.

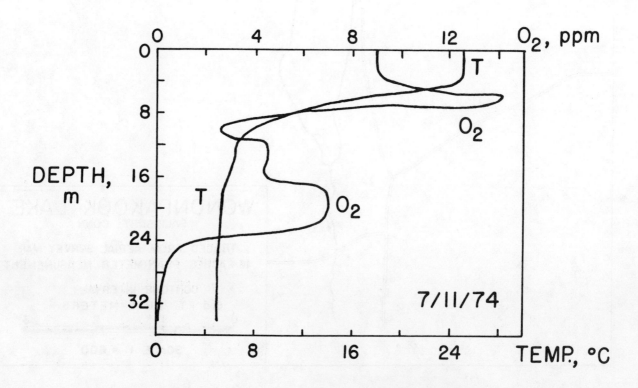

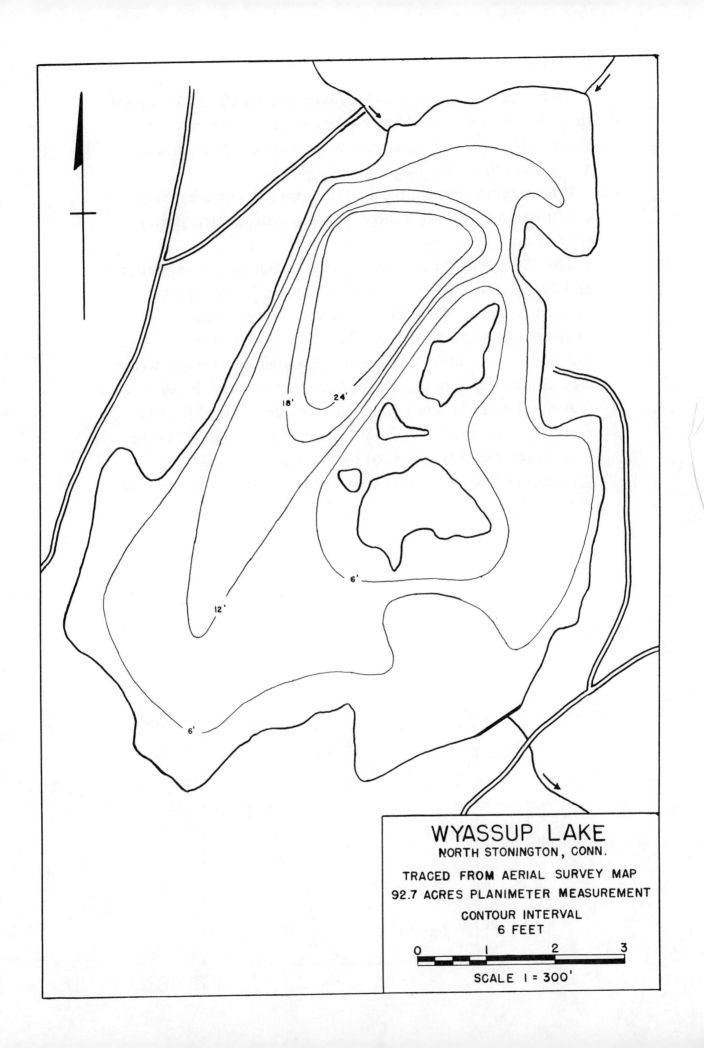

ZOAR LAKE

Zoar Lake is located in New Haven and Fairfield Counties in the towns of Newtown, Monroe, Oxford, and Southbury. It has a surface area of 975 acres, a maximum depth of 75 feet and an average depth of 24.6 feet.

It is an impoundment on the Housatonic River created by the construction in 1919 of a large dam for hydroelectric power generation.

The 986,240-acre watershed is 8.1% urban, 16.7% agricultural and 75.2% wooded or wet. The Housatonic River drains a substantial portion of Connecticut and smaller portions of Massachusetts and New York. The discharge from Lake Lillinonah is the most significant hydrologic contributor to the lake. Shoal areas are rare due to the steep slope of the shore.

A water and nutrient budget has been determined for Lake Zoar by Aylor and Frink (1980). Distribution of PCBs in the bottom sediments is described by Frink, et al. (1982).

Access is through a State boat launching area in Kettletown State Park.

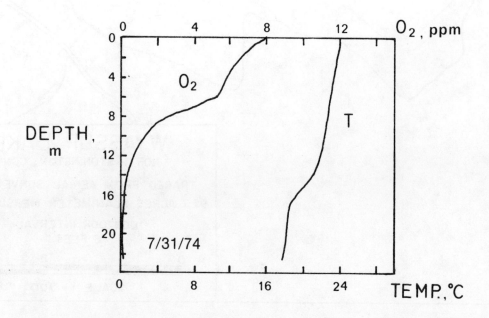

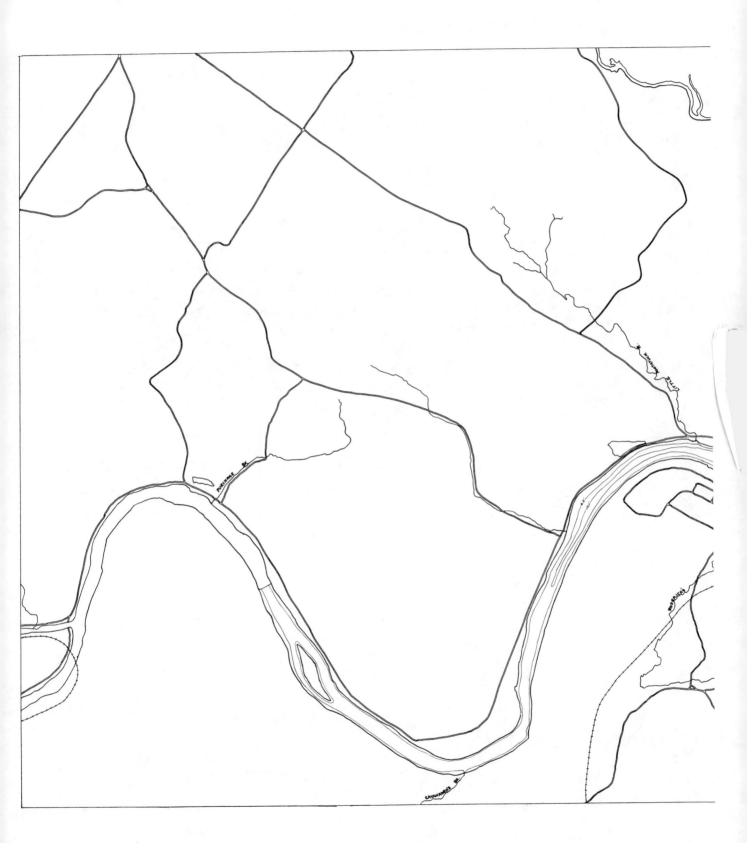

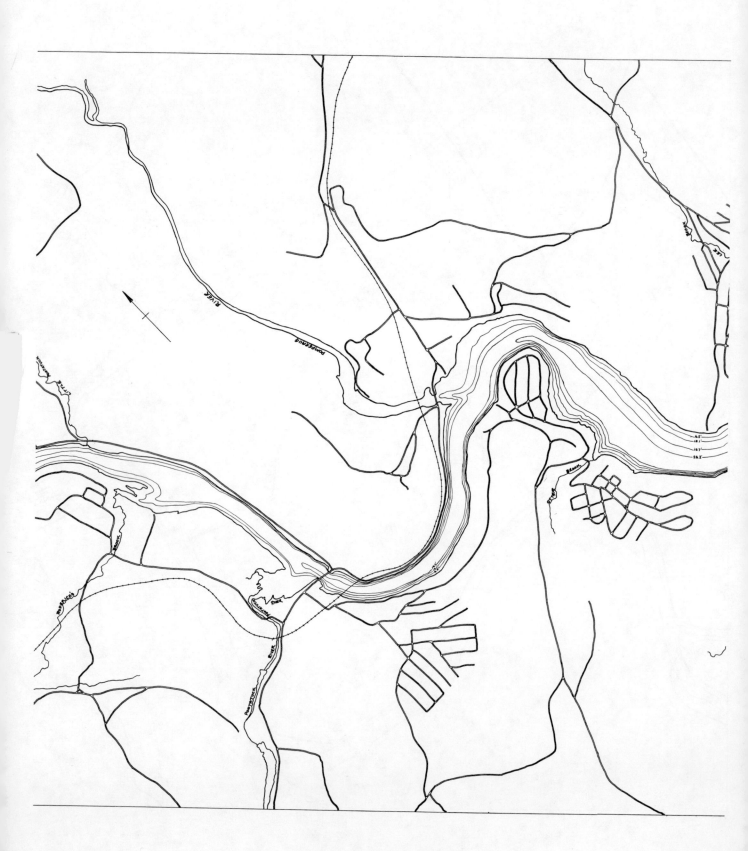

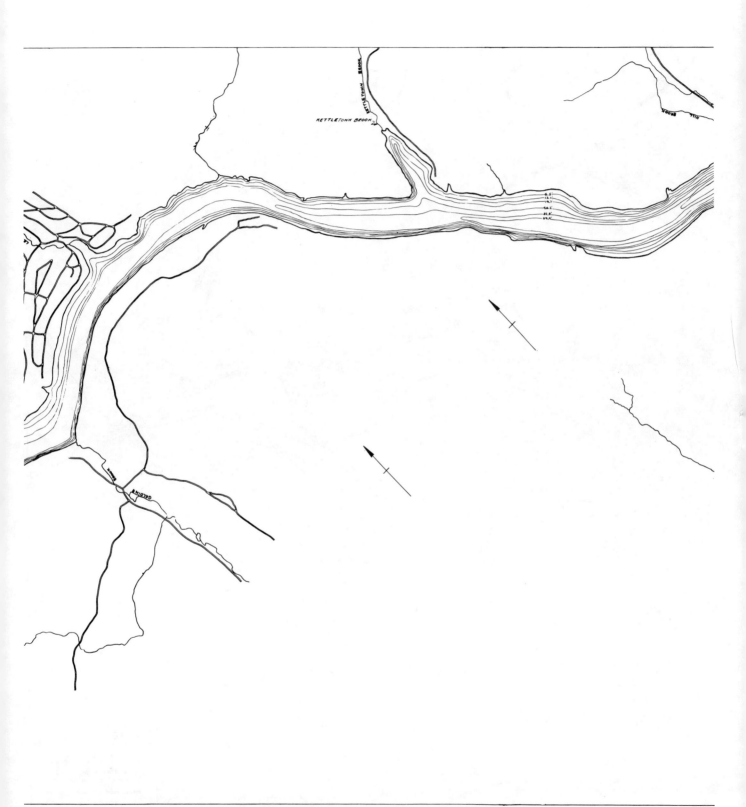

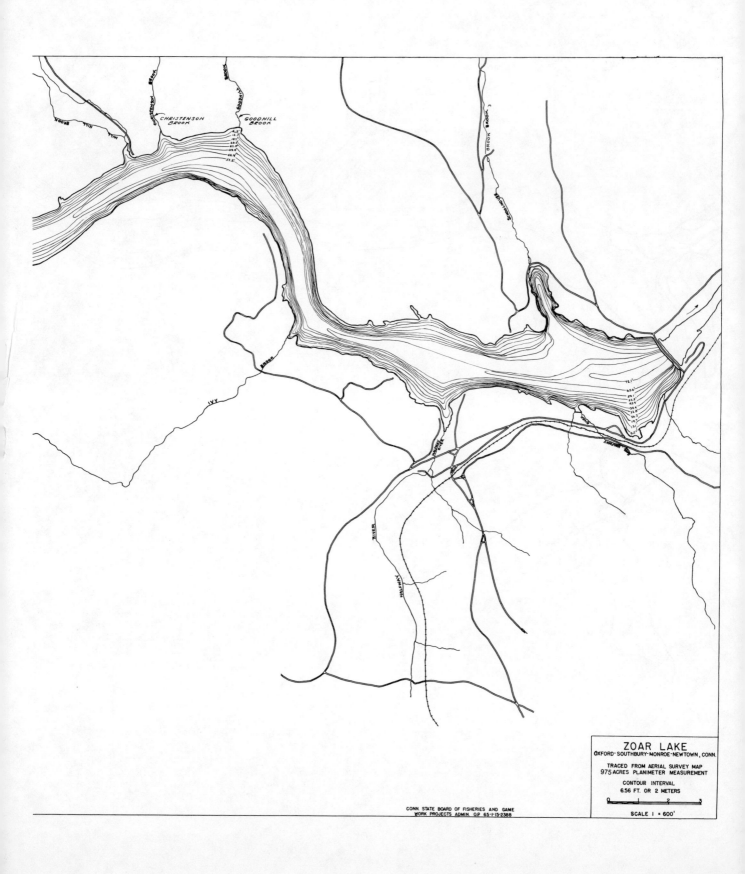